Biomimicry and Sustainable Building Performance

This book on biomimicry assessment tools studies the concepts of sustainability, sustainable construction practices, and the evaluation categories that constitute a sustainability assessment tool.

By exploring and drawing lessons from biomimicry principles, the book provides a nature-inspired assessment tool to aid and guide the sustainable transformation of the built environment. The book encapsulates the attributes of the conceptualised biomimicry assessment tool, which is aimed at helping practitioners, regulatory bodies, and governmental and non-governmental agencies in greening the built environment. Owing to the dire need for country-specific and tailor-made tools that address developing countries' needs, this book serves as a practical reference and theoretical springboard for the development of sustainability assessment tools for the built environment. Furthermore, the book serves as a guide in navigating the path towards achieving the greening agendas of the built environment and other sectors and seeks to align the new biomimicry assessment tool with the UN Sustainable Development Goals (SDGs).

It is important reading for academics, professionals and advanced students in the built environment, engineering, and sustainable development.

Olusegun Aanuoluwapo Oguntona is a Senior Lecturer in the Department of Built Environment, Walter Sisulu University, South Africa. He has a PhD from the University of Johannesburg and has previously worked in industry as an Architect and Project Manager. He is a member of the Chartered Institute of Building (CIOB), Project Management South Africa (PMSA), Green Building Council Nigeria (GBCN), and the South African Council for the Project and Construction Management Professions (SACPCMP). His research interests are in biomimicry, sustainable construction, green architecture, green building, biomimetics and innovative technologies, and sustainable construction practices.

Clinton Ohis Aigbavboa is a Professor in the Department of Construction Management and Quantity Surveying and Director of cidb Centre of Excellence & Sustainable Human Settlement and Construction Research Centre, University of Johannesburg, South Africa. Before entering academia, he was involved as a quantity surveyor on several infrastructural projects in Nigeria and South Africa. He has published several research papers and more than ten research books in housing, construction and engineering management and research methodology for construction students. He is the editor-in-chief of the *Journal of Construction Project Management and Innovation.*

Routledge Research Collections for Construction in Developing Countries
Series Editors: Clinton Aigbavboa, Wellington Thwala, Chimay Anumba, David Edwards

Moving the Construction Safety Climate Forward in Developing Countries
Sharon Jatau, Fidelis Emuze and John Smallwood

Unpacking the Decent Work Agenda in Construction Operations for Developing Countries
Tirivavi Moyo, Gerrit Crafford, Fidelis Emuze

An Integrated Infrastructure Delivery Model for Developing Economies: Planning and Delivery Management Attributes
Rembuluwani Bethuel Netshiswinzhe, Clinton Aigbavboa and Didibhuku Wellington Thwala

A Roadmap for the Uptake of Cyber-Physical Systems for Facilities Management
Matthew Ikuabe, Clinton Aigbavboa, Chimay Anumba and Ayodeji Oke

A Building Information Modelling Maturity Model for Developing Countries
Samuel Adekunle, Clinton Aigbavboa, Obuks Ejohwomu, Didibhuku Wellington Thwala, Mahamadu Abdul-Majeed

Training-Within-Industry Job Programs for Improved Construction Safety
Lesiba Mollo, Fidelis Emuze and John Smallwood

Biomimicry and Sustainable Building Performance
A Nature-inspired Sustainability Guide for the Built Environment
Olusegun Oguntona and Clinton Aigbavboa

Biomimicry and Sustainable Building Performance

A Nature-inspired Sustainability Guide for the Built Environment

Olusegun Aanuoluwapo Oguntona and Clinton Ohis Aigbavboa

Routledge
Taylor & Francis Group

LONDON AND NEW YORK

First published 2024
by Routledge
4 Park Square, Milton Park, Abingdon, Oxon OX14 4RN

and by Routledge
605 Third Avenue, New York, NY 10158

Routledge is an imprint of the Taylor & Francis Group, an informa business

British Library Cataloguing-in-Publication Data
A catalogue record for this book is available from the British Library

ISBN: 978-1-032-53896-9 (hbk)
ISBN: 978-1-032-54256-0 (pbk)
ISBN: 978-1-003-41596-1 (ebk)

DOI: 10.1201/9781003415961

Typeset in Times New Roman
by KnowledgeWorks Global Ltd.

This book is dedicated to God, humanity, and nature.

Also, to my dearest son Enoch Oluwademilade Oguntona, this book is dedicated to you with all my heart. You have been a source of joy and inspiration.

Contents

Tables

Figures

Preface

The continuous and increasing adverse impact of the built environment on both the human and natural environment is alarming. These impacts include the generation of substantial waste, the release of volatile organic compounds (VOCs) and greenhouse gases (GHGs), excessive consumption of energy, water, and natural resources, as well as various forms of pollution such as air, noise, waste, and water pollution. This has led to the global call for various concepts and practices to minimise and address the severity of these impacts. A potent concept that is identified and widely believed will comprehensively curb this menace is green building (GB). Also referred to as sustainable building, GB is premised on the principles of sustainable construction and is found to pose minimal environmental impacts, consume less energy, and be resource-efficient. The overarching goal of GB is to mitigate and address these negative consequences. Despite the global push for the adoption and implementation of GB practices in the construction industry (CI), it is evident that numerous detrimental environmental impacts persist within the industry's processes and activities.

The transition towards GB practices has been sluggish, primarily due to the continued specification and use of traditional construction materials and technologies, as well as the engagement in unsustainable construction processes and activities by industry professionals and stakeholders. Additional factors contributing to this slow progress include resistance to change and the adoption of new practices, the persistence of misconceptions surrounding green practices, products, and services, instances of greenwashing, limited awareness and knowledge of green materials and technologies, and the absence of comprehensive green labelling and evaluation tools, among other challenges. Consequently, there is an urgent need to overcome these barriers and facilitate more rapid and widespread adoption of GB practices, especially with the aid of green building rating tools (GBRTs).

The primary objective of GBRTs is to assess and certify building projects while offering valuable information to construction stakeholders and consumers about a building's sustainable features. These rating systems aim to evaluate a building project's entire life cycle, including critical aspects like global warming potential, emissions of air, water, and land, resource utilisation (both absolute and relative), embodied energy, energy consumption, and waste generation, among others. However, a comprehensive examination of the existing GB rating systems, whether they focus on a single attribute or multiple attributes, reveals a predominant emphasis on the environmental dimension of sustainability, with limited attention given to the social and economic aspects. This observation indicates that rating systems falling under the multiple-attributes category fail to encompass the entirety of sustainability dimensions, thereby justifying their classification as environmental rating tools. While popular environmental rating systems like BREEAM and LEED have undergone constant upgrades and revisions, a tool that thoroughly considers all three sustainability pillars is imperative. There is, therefore, a pressing need for an up-to-date sustainability assessment tool that effectively addresses contemporary sustainability requirements in an integrated manner. There is, therefore, a dire need for an objective, easy-to-use, verifiable, voluntary, affordable, and broad-scoped assessment tool that addresses sustainability issues in its entirety.

This book made a significant contribution to the body of knowledge as it explores existing GBRTs utilised globally. The book conceptualises a biomimicry sustainability assessment tool that will enable the evaluation of buildings against research-based and valid criteria in the quest to ensure the sustainability performance of buildings. Theoretically, this book developed a sustainability assessment tool using ten latent constructs with the inclusion of three new constructs: inclusivity and equality, collaboration, and resilience. It is noteworthy that previous GBRTs considered other constructs without the inclusion of three new ones adopted for this study. A comprehensive distillation of biomimicry (nature) principles and sustainable development goals informed the new constructs thereby ensuring the novelty of this tool.

Considering the vast depth of innovative ideas in biomimicry and the greening agenda of the built environment, this book can be adapted to the development of single and multiple-attribute sustainability assessment tools. This book will also be a useful material and guide in defining the corporate sustainability goals of multinationals, established, and small-scaled companies. Also, this book will greatly benefit the major players in the green market space to avoid and tackle the sins of greenwashing thereby increasing consumer's confidence in green products and services. The book can also serve as a reference material for researchers with an interest in sustainable development within the built environment and other sectors. Also, this book can be adopted to aid the development of modules and courses that addresses sustainability in the built environment programmes. This book will be of interest to researchers and policymakers in the built environment, sustainability proponents, practising construction professionals, green building agencies and councils, built environment professionals, regulatory bodies, higher educational institutions, and government parastatals responsible for the adoption, and implementation of sustainable development. Hence, the authors confirm that the text utilised in this book reflects original scholarly work and, where necessary, materials that the book benefited from were cited and referenced appropriately.

Olusegun Aanuoluwapo OGUNTONA
Clinton Ohis AIGBAVBOA
July 2023

Acknowledgements

We sincerely appreciate and acknowledge all those who made significant contributions, inputs, and suggestions to this book project. We especially acknowledge biomimicry proponents around the world, the Biomimicry Institute, Biomimicry 3.8, BiomimicrySA, the cidb Centre of Excellence and the Department of Construction Management and Quantity Surveying, University of Johannesburg, South Africa, and the Directorate of Research Development and Innovation, Walter Sisulu University, South Africa for providing the opportunity to successfully embark and complete this book project.

About the Authors

Olusegun Aanuoluwapo OGUNTONA is a sustainability proponent, architect, and construction project manager by profession. He has a BTech (Hons) in Architecture from Ladoke Akintola University of Technology, Nigeria. Thereafter, he practised under notable architectural firms in Nigeria as an Architect between 2009 and 2015. He obtained his master's degree in Construction Management (with distinction) from the University of Johannesburg, South Africa. He was the recipient of the 2017 Chancellor's Medal for the Most Meritorious Master's Study in the Faculty of Engineering and the Built Environment, University of Johannesburg, South Africa. He thereafter obtained his PhD in Construction Management from the University of Johannesburg. Dr. Oguntona is currently a Senior Lecturer at the Department of Built Environment, Faculty of Engineering and Technology, Walter Sisulu University, South Africa. He currently serves as the Chairperson: Department of Built Environment Research and Higher Degrees Committee, Chairperson: Faculty of Engineering and Technology Research and Higher Degrees Committee, Member: Senate Research and Higher Degrees Committee of Walter Sisulu University, South Africa. His research interests are in biomimicry, sustainable construction, green architecture, green building, biomimetics and innovative technologies, and sustainable construction practices. He also serves as a regular reviewer for various Taylor & Francis, Springer, MDPI, Elsevier, and Emerald built environment journals. He is a member of the Chartered Institute of Building (CIOB), Project Management South Africa (PMSA), and the South African Council for the Project and Construction Management Professions (SACPCMP).

Clinton Ohis AIGBAVBOA is a full Professor at the Department of Construction Management and Quantity Surveying, University of Johannesburg, South Africa. Before joining academia, he was involved as a quantity surveyor on several infrastructural projects, both in Nigeria and South Africa. Prof. Aigbavboa is the immediate past Vice Dean of the Faculty of Engineering and the Built Environment at the University of Johannesburg, South Africa. He has extensive knowledge in practice, research, training, and teaching. He is currently the Director of the Construction Industry Development Board (cidb) Centre of Excellence and the Sustainable Human Settlement and Construction Research Centre at the University of Johannesburg. He is also the author of over ten research books that were published by Springer Nature and CRC Press. He is currently the editor of the Journal of Construction Project Management and Innovation (accredited by the DHET) and has received national and international recognition in his field of research. He is rated by the National Research Foundation in South Africa.

List of Abbreviations

AHP	Analytic Hierarchy Process
AM	Additive Manufacturing
BIM	Building Information Modelling
BioSAT	Biomimicry Sustainability Assessment Tool
BLP	Biomimicry Life's Principle
BRE	Building Research Establishment
BREEAM	Building Research Establishment Environmental Assessment Method
BT	Blockchain Technology
CASBEE	Comprehensive Assessment System for Built Environment Efficiency
CHES	Coupled Human Environment Systems
CI	Construction Industry
CM	Cost Management
CO	Collaboration
CO$_2$	Carbon dioxide
CSR	Corporate Social Responsibility
DM	Decision-making
EDGE	Excellence in Design for Greater Efficiencies
EE	Energy Efficiency
ES	Ecosystem Services
EV	Eigenvector
FIT	Fully Integrated Thinking
GBCSA	Green Building Council of South Africa
GBRT	Green Building Rating Tool
GDP	Gross Domestic Product
GE	Gender Equality
GHG	Greenhouse gas
IAPS	Invasive alien plant species
IBS	Industrialised Building System
ICT	Information Communication Technology
IE	Inclusivity and Equality

ISO	International Organisation for Standardisation
LC	Lean construction
LEED	Leadership in Energy and Environmental Design
MDG	Millenium Development Goals
ME	Materials efficiency
NC	New Construction
NDP	National Development Plan
NGO	Non-governmental organisation
NZEB	Net zero energy building
QUAL	Qualitative
QUAN	Quantitative
SA	South Africa
SAT	Sustainability Assessment Tool
SB	Sustainable building
SBAT	Sustainable building assessment tool
SBTool	Sustainable building tool
SC	Sustainable Construction
SD	Sustainable Development
SDG	Sustainable Development Goals
SE	Site ecology
STEM	Science, technology, engineering and mathematics
TERI	The Energy and Resources Institute
UC	User comfort
UK	United Kingdom
UN	United Nations
UNEP	United Nations Environment Programme
USGBC	United States Green Building Council
VE	Value engineering
VOC	Volatile organic compound
WCED	World Commission on Environment and Development
WE	Water efficiency
WGBC	World Green Building Council
WLC	Whole life costing
WP	Waste and pollution

Part I

Background Information of the Book

1 General Introduction

Introduction

The construction industry (CI) plays an important role in enhancing the quality of life, providing basic amenities and infrastructure and contributing to the economic development of a country. In the process of making these beneficial contributions to the human environment, the CI has become that sector that continually disrupts the peace and harmony existing in the natural world. Since the natural environment supports human life, the CI has attracted global attention, coupled with heavy criticism due to the numerous ongoing detrimental effects it has on the environment. Despite the positives and the beneficial characteristics of the industry, the negative environmental impacts are on the increase and overwhelming. Consumption of large amounts of natural resources, significant sources of global carbon emissions, waste generation, and a high rate of energy consumption which in turn results in land, air, and water pollution (Madurwar et al., 2013; Wu et al., 2015) are few of the issues attributed to construction processes and activities. Hence, the need to embrace sustainable building practices or sustainable construction is identified as the most potent panacea to the growing environmental concerns of the CI (Doan et al., 2017).

There is an increased surge of interest in sustainability concepts which is traceable to the global attention and efforts towards addressing the environmental and other adverse impacts of the CI (Kubba, 2016). The terms 'sustainable construction' (SC), 'sustainable building' (SB), 'green building' (GB), 'high-performance', and 'environmentally friendly', amongst others, are often used interchangeably to describe the transition of the CI towards achieving sustainability (Ahn et al., 2013; Kibert, 2013; Yılmaz & Bakış, 2015). However, these terms define the same level of responsiveness towards addressing the environmental impacts of the CI. According to Haapio and Viitaniemi (2008), Li et al. (2016), and Doan et al. (2017), GB can be described as 'providing people with healthy, applicable, efficient space and natural harmonious architecture with the maximum savings on resources (energy, land, water, materials), protection for the environment and reduced pollution throughout its whole lifecycle'. The focal point of GB is to proliferate the efficient use of water, energy, and materials while minimising its footprint on the environment, human health, and safety (Dodo et al., 2015). It is premised on the provision

DOI: 10.1201/9781003415961-2

of a socially, technically, and economically pleasing and blooming human environment with minimal or no harm to the natural environment (Suzer, 2015). However, despite the global recognition of the importance of GB, its demand, adoption, and implementation are still in its infancy owing to a lack of knowledge, awareness, eco-friendly materials, and sustainability measurement tools (Hankinson & Breytenbach, 2012; Serpell et al., 2013; AlSanad, 2015; Ametepey et al., 2015). Hence, the terms 'sustainable construction' (SC), 'sustainable building' (SB), 'green building' (GB), 'high-performance', and 'environmentally friendly', amongst others, are referred to as GB or SB in this book.

According to Said and Harputlugil (2019), there is a consensus among scientists and researchers around the globe that one of the major ways of addressing the negative environmental impacts of the CI is by ensuring buildings (existing and new developments) are sustainable in all their ramifications. However, for a building project to be referred to as sustainable or green, a few standards such as materials, energy, and water efficiency are to be met. A critical assessment of a building project against certain parameters and criteria will determine how green it is or what its sustainability level is. This led to the development of several assessment tools to help in the actualisation of GBs. According to Cohen et al. (2017) and Doan et al. (2017), GB assessment tools are established by organisations, bodies, and authorities to reduce greenhouse gas (GHG) emissions, control pollution, minimise the consumption of natural resources, and ensure water and energy efficiency. In literature, GB assessment tools are also referred to as sustainability certification schemes, Sustainable design, GB rating programmes, sustainability assessment tools, GB rating systems, environmental assessment tools, and environmental labels, amongst others (Castro-Lacouture et al., 2009; Fuerst & McAllister, 2011; Thilakaratne & Lew, 2011; Altomonte & Schiavon, 2013; CSI, 2013; Kubba, 2016; Siebers et al., 2016). About 600 sustainability assessment tools in the form of standards, guidelines, protocols, eco-labels, and systems have been developed and are currently in existence (Li et al., 2016; Doan et al., 2017). According to Zuo and Zhao (2014) and Li et al. (2017), prominent and well-known assessment tools include Leadership in Energy and Environmental Design (LEED), Building Research Establishment Environmental Assessment Method (BREEAM), Comprehensive Assessment System for Built Environment Efficiency (CASBEE, Japan), Green Mark Scheme (Singapore), Evaluation Standard for Green Building (ESGB) or Green Building Label (GBL) of China, Deutche Gesellschaft fur Nachhaltiges Bauen (DGNB) of Germany, Building Environmental Assessment Method Plus (BEAM Plus), Hong Kong Building Environmental Assessment Method (HK BEAM), Green Building Index (Malaysia), EcoProfile (Norway), Green Building Council of Australia Green Star (GBCA, Australia), Pearl Rating System for Estidama (Abu Dhabi Urban Planning Council), and the international Sustainable Tool (SB Tool) previously known as the GB Tool. Ali and Al Nsairat (2009) also listed EcoQuantum (Netherlands), KCL Eco (Finland), Bees (USA), and Beat (Denmark) as other available assessment tools. However, the most used and widely acceptable among these assessment tools are LEED and BREEAM (Yılmaz & Bakış, 2015), owing to their wide recognition, timely renewals, and constant incorporation of the latest regulations, technologies, and experience in practice (Liu et al., 2019). By updating existing versions from time to time, BREEAM and LEED assessment tools can provide eco-labelling mechanisms and GB evaluations that are effective and reliable.

The principal aim of GB rating systems can be summarised into two, namely promote sustainable (high-performance) buildings and help create and increase the demand for sustainable construction (Kubba, 2016). These assessment tools are voluntary, type-specific, and country/region-specific, and developed for different types of buildings (i.e., residential, industrial, commercial) and urban developments which are assessed differently and with different criteria. The adoption and use of these tools aim to evaluate and provide verifiable sustainability ratings (Kim et al., 2013) for building projects. With a range of gains attributed to GBs such as incentives, tax reliefs, subsidies, reduced utility cost, higher occupancy rates, low-risk premiums, and reputational benefits (Fuerst & McAllister, 2011), GB rating systems play a prominent role in achieving sustainability in the CI and advancing the growth of GBs. Also, GB rating systems offer a thorough evaluation of buildings against a broad spectrum of indicators or criteria. Upon a satisfactory outcome, a building can be tagged as sustainable or green (Jawali & Fernández-Solís, 2008).

The CI industry employs two categories of tools in assessing the sustainability performance of building projects, namely rating and assessment tools. Rating tools aid the determination of a building's performance level in quantitative systems while assessment tools provide performance benchmarks in quantitative forms for design alternatives (Jawali & Fernández-Solís, 2008). From the literature, sustainability assessment tools are grouped into two, namely scores or criteria-based systems and life cycle assessment methodology (LCA). The criteria-based tools are those that assign point values to selected criteria (i.e., BREEAM, LEED) while the LCA (i.e., EcoQuantum, KCL Eco) uses different weighing techniques based on a different rationale for evaluation to be deployed for the selection of building design, material, and local utility options (Ali & Al Nsairat, 2009). Owing to the complexities in the greening of infrastructural projects, these rating tools address different aspects and categories. For example, the GBCA Green Star rating tools tackle nine aspects, namely material, emissions, transport, energy, indoor environmental quality, management, water, land use and ecology, and innovation (Zuo & Zhao, 2014). The LEED rating tool emphasises sustainable sites, water efficiency, energy and atmosphere, materials and resources, indoor environmental quality, innovation in design, and regional priority (recently added) categories (Poveda & Lipsett, 2011; Lee, 2013) while the Malaysian Green Building Index, derived from the Australian Green Star System and Green Mark Singapore, addresses six vital benchmarks, namely material and resources, indoor environmental quality, water efficiency, energy efficiency, innovation, and sustainable site planning and management (Dodo et al., 2015). The BREEAM rating tool is structured into ten (10) distinct issues, namely health and well-being, materials, pollution, transport, energy, management, water, land use and ecology, waste, and innovation (Lee, 2013; Doan et al., 2017; Li et al., 2017). However, a comprehensive review of these assessment tools reveals that they strongly tend towards achieving the environmental arm of the sustainability pillar. Hence, the need for an assessment and rating tool that encompasses all three pillars of sustainability more holistically is a gap this book aims to fill.

Biomimicry, the identification, and integration of predominantly sustainable processes and strategies in nature to proffer solutions to human challenges (Goss, 2009), has become a popular sustainability trend in recent years. The biomimicry idea is premised on the notion that nature is an infinite provenance of knowledge, resulting in the birthing of innovative solutions to present-day human challenges (Nychka & Chen, 2012). Biomimicry principles (also known as nature principles) are those factors guiding the sustainable existence of natural organisms on earth. These principles serve as guides towards a better understanding of biomimicry and its application to achieve the overarching goal of holistic sustainability. As a result of decades of intense study, a series of nature's principles, strategies, and laws have been discovered, namely, nature runs on sunlight, nature fits form to function, nature rewards cooperation, nature taps the power of limit, nature banks on diversity, nature recycles everything, nature demands local expertise, nature uses only the energy it needs, and nature curbs excesses from within (Benyus, 1997; Berkebile & McLennan, 2004).

Based on these principles which are exhibited by nature, biomimicry life's principles (BLPs) have evolved. As reworked by the Biomimicry Institute, now known as Biomimicry 3.8, BLPs are segmented into six broad categories, namely evolve to survive, be locally attuned and responsive, use life-friendly chemistry, adapt to changing conditions, integrate development with growth, and be resource efficient (Ariffin & Gad, 2015). Since it came into the limelight, biomimicry has now become the provider of numerous outstanding innovations in the areas of waste reuse, energy engineering, and most significantly, materials and technologies with sustainable credentials. Many researchers have investigated biomimetic materials (also known as nature-inspired or bio-inspired), especially in the building and CI, and found them to possess sustainable attributes compared to traditional ones (Molinaro et al., 2016). Today, the application of biomimicry now spans various fields of human endeavour emphasising the attainment of sustainability in a complete form (Iouguina, 2013). An example is the content analysis by Ariffin and Gad (2017) aimed at integrating BLPs into the Green Building Index of Malaysia to achieve a higher level of environmental sustainability. Hence, this book seeks to present and infuse the integrated set of values exhibited by nature to survive, live, and thrive over the years BLPs in tandem with the Sustainable Development Goals (SDGs) for the conceptualised Biomimicry Sustainability Assessment Tool (BioSAT) for building projects in developing countries.

Aim and Objectives of the Book

Despite the global call for the adoption and implementation of GB in the CI, processes, and activities within the industry still come with numerous adverse environmental impacts. Production of significant amounts of waste, emission of volatile organic compounds (VOCs) and greenhouse gases (GHGs), consumption of huge amounts of energy, water, and natural resources, and pollution (air, noise, waste, and water) are a few of the negative impacts of the CI that GB aims to address. However, the slow transition towards GB practices is traceable to the continuous specification and use of traditional construction materials and technologies and engaging in unsustainable construction processes and activities by professionals and stakeholders in the industry. Other factors include resistance to change in embracing new practices; perceived myths on adopting green practices, products, and services; greenwashing; lack of awareness and knowledge of green materials and technologies, and lack of comprehensive green labelling and evaluation tools, amongst many other factors.

With the growing number of GB rating systems and their acceptance in more countries around the world, the CI is yet to experience a full transformation towards sustainability. It is also found that prominent GB rating systems such as the BREEAM and LEED mainly address the environmental concerns of the CI while being less concerned with the social and economic considerations. The study of Awadh (2017) on the BREEAM, LEED, Pearl Rating System for Estidama, and Global Sustainability Assessment System (GSAS) revealed the concentration and focus of these rating tools on the environmental aspect of sustainability. As affirmed by Krajangsri and Pongpeng (2018), there is a paucity of GB rating tools that can be employed in effectively evaluating the sustainability (economic, social, and environmental conditions) of construction projects. Considering the vast number of GB rating tools that are actually 'environmental rating tools', the goal of having truly 'sustainability assessment tools' that will incorporate the three pillars of sustainability seems challenging (Haapio & Viitaniemi, 2008; Goh, 2014), a feat this book seeks to achieve.

Considering the scope of GB rating systems, they are categorised to be either narrow- or broad-scope systems (Larsson, 2004). The narrow-scope systems focus on single issues such as energy and emissions in the Australian Building Greenhouse Rating System. On the other hand, broad-scope systems emphasised addressing broader issues such as indoor air quality, and employment generation among others, of which most of the existing GB rating systems fall short. Another major challenge posed by the present-day GB rating systems is the relatively high cost of assessment charged by the certification body and the required certified/accredited assessor. Significant fees are required to be paid in preparing for certification (gathering and preparing the information/documents needed to apply) while the assessor's fees are also a substantial amount.

Whether performing an external or self-assessment, the rigour and complexities involved in a full-scale assessment have proven to be another issue identified with the existing GB rating systems. Experienced and duly accredited assessors/ professionals are required for the assessment process which is cumbersome and rigorous coupled with many technicalities. Each of the building and construction categories addressed by the GB rating systems requires an expert assessor who has been trained and possesses a considerable understanding of the environmental issues involved. Considering the LEED rating system, for example, an accredited professional (AP) for the Building Design + Construction (LEED AP BD+C) is not trained to handle the assessment of the Operations + Maintenance category which requires the services of a LEED AP O+M.

A critical analysis of the existing GB rating systems also reveals their inability to address and fully align with the SDGs – a United Nations (UN) 2030 Agenda for Sustainable Development plan of action for people, the planet, and prosperity. The SDGs aim to achieve a synergy of the three dimensions of sustainable development (SD) (environmental, economic, and social) in a balanced and integrated way. There is, therefore, a dire need for an objective, easy-to-use, verifiable, voluntary, affordable, and broad-scoped GB rating system that addresses the issue of sustainability in its entirety. Therefore, the objective of this book was to conceptualise a BioSAT for optimising and accelerating the sustainability drive of the built environment. With the conceptualised BioSAT, environmental concerns together with the economic and social issues of sustainability will receive equal attention in achieving 'truly' green (sustainable) building projects in the CI.

Significant Contributions of the Book

The essence of GB rating systems is to evaluate and certify building projects while providing construction stakeholders and consumers with information regarding the sustainable attributes such a building possesses. By evaluation, these rating systems are meant to provide an inventory of a building project's life cycle assessment, showing how it has addressed issues such as global warming potential, air, water and land emission, resource use (weighed and absolute), embodied energy and energy consumption, and waste generation, amongst many others. However, a careful assessment of the array of existing GB rating systems (single or multiple attributes) used in various parts of the world reveals partial or whole concentration on the environmental pillar of sustainability while neglecting the social and economic aspects. It can be inferred that those rating systems that fall into the multiple-attributes category fail to address the totality of the sustainability dimensions, hence justifying their perception and classification as environmental rating tools. While the most popular and widely used environmental rating tools such as BREEAM and LEED have constantly subscribed to upgrades and revisions, the few such as the Sustainable Building Assessment Tool (SBAT) of South Africa that encompass the three tiers of sustainability remain stagnant and seem obsolete. Hence, the need for an up-to-date sustainability assessment tool that addresses the present-day building sustainability needs in an integrated manner. This book, therefore, provides the pathway to defining the necessary categories to assess the sustainability performance of buildings.

This book evaluated existing GB rating systems to identify and establish the requisite attributes of sustainability assessment tools coupled with biomimicry principles (nature-inspired attributes) and aligned with the SDGs to conceptualise a holistic assessment tool that reflects an integration of the three dimensions of sustainability. Despite the novelty and the vast potential embodied within the biomimicry concept, its uptake, and implementation in the CI domain is still new, a gap this book identified and addressed. As a field of study that provides sustainable lessons from nature to proffer solutions to the challenges facing humanity, biomimicry's overarching goal is holistic sustainability. Thus, this book focused on projecting the extracted sustainability attributes that are inherent in natural ecosystems in developing the conceptualised tool. While existing and renowned assessment tools as well as relevant stakeholders mainly focus on environmental issues of sustainability, cogent factors such as resilience, collaboration, inclusivity, and equality are partly ignored. By addressing these important factors collectively and in tandem with the SDGs, this book provides an excellent platform upon which tools for specific buildings or product types can be developed, enhanced, and further adapted for use.

The conceptualised tool in this book is essential and timely considering the vast records of greenwashing and related issues traceable to the global clamour for sustainable buildings and products. The findings from this book are significant to sustainability proponents, government, and non-governmental agencies as they provide the necessary foundation upon which policies and frameworks for SD can be built. As the first book to introduce a biomimicry assessment tool by providing a nature-inspired sustainability guide for the built environment, the uniqueness of this book is that it provides an adaptable, comprehensible, and all-encompassing criterion for determining the sustainability performance of building projects. Despite the numerous assessment tools in existence, it is important to have a tool that has considerations for the peculiarity of the African continent and other developing countries. Hence, this book contains updates and relevant content that speaks to the distinct nature and needs of Africa and other developing countries by integrating the core tenets of SD and the SDGs of the United Nations. This book provides significant recommendations and insights for stakeholders in their efforts to realise the greening agendas of the built environment and other sectors of the economy.

In summary, the conceptualised tool will be beneficial to the professionals and other stakeholders in the CI to advance the greening agenda of the industry. By providing a reliable, robust, credible, cost-effective, easy-to-use, and objective sustainability assessment tool option in the CI, the BioSAT will offer an avenue for the proliferation of green (sustainable) buildings while creating a sustainability consciousness among professionals and other stakeholders in the CI. Also, the Bio-SAT forms an explicit foundation upon which further research can be conducted to adapt its use to specific building types, countries, and regions. The outcome of this research has the potential to provide evidential bases and information that will inform the development of single-attribute sustainability rating systems.

Structure of the Book

For easy navigation, this book is organised into four parts consisting of 11 chapters in all. Part I is on the background information of the book. This chapter presents important details about the book such as the aim and objectives, significance, and structure of the book. Part II of the book consists of four chapters tailored towards providing an in-depth understanding of sustainability trends in construction. Chapter 2 explores the sustainability nexus in the CI by providing an overview of the state of the industry, the sustainability concept, SD, and the United Nations'

SDGs. Chapter 3 presents the incorporation of sustainability in construction by examining the various trends, practices, and concepts for optimising sustainability in the sector. Chapter 4 is developed to present a comprehensive understanding of the green building concept. The overview, barriers, benefits, and drivers of green building were discussed. Chapter 5 introduces the biomimicry paradigm which entails nature inspiration and emulation. The evolution of biomimicry terminologies, definitions, dimensions, characteristics, approaches, and principles of biomimicry were discussed. Part III is focused on sustainability assessment and rating in the construction sector and consists of three chapters. Chapter 6 examines sustainability assessment tools in the CI to highlight the core attributes and taxonomy of sustainability assessment tools. Chapter 7 presents the various notable and widely used sustainability tools and methods for building evaluation. The core tools, systems, and methods for green building evaluation were mainly discussed. Chapter 8 focuses on presenting the green initiatives, drives, and assessment tools for buildings in South Africa. The green building collaborative networks in South Africa are also discussed. Part IV focuses on the conceptual understanding of BioSAT for the CI and consists of three chapters. Chapter 9 examines the gaps in sustainability assessment tools research that are best fit for the conceptualised framework. Chapter 10 explores the various dimensions, attributes, and sub-attributes that form the conceptualised sustainability assessment tool. Chapter 11 explores the conceptualised model in a developing country using expert judgements from the pairwise comparison questionnaire. The significance of the dimensions, categories, and criteria is assessed using a developing country's case study.

Summary

This chapter introduces the fundamentals of this research book with an emphasis on SD in the CI and sustainability assessment tools for buildings. The various constituting chapters help to support and assert the relevance of this book to the built environment of now and the future. The main objective of this book is presented which is to develop a BioSAT for the CI. The tool aims to ascertain the influence of energy efficiency, water efficiency, materials efficiency, waste and pollution, site ecology, user comfort, cost management, collaboration, resilience, and inclusivity and equality in predicting and determining the sustainability performance of a building. This book addresses the gaps in sustainability assessment in the built environment to ensure a tool that evaluates all dimensions of SD is developed and activated. The next chapter presents the concept of sustainability, SD, and the SDGs which provides part of the key principles required in developing the sustainability assessment tool.

References

Ahn, Y. H., Pearce, A. R., Wang, Y., & Wang, G. (2013). Drivers and barriers of sustainable design and construction: The perception of green building experience. *International Journal of Sustainable Building Technology and Urban Development, 4*(1), 35–45.

Ali, H. H., & Al Nsairat, S. F. (2009). Developing a green building assessment tool for developing countries – Case of Jordan. *Building and Environment, 44*(5), 1053–1064.

AlSanad, S. (2015). Awareness, drivers, actions, and barriers of sustainable construction in Kuwait. *Procedia Engineering, 118*, 969–983.

Altomonte, S., & Schiavon, S. (2013). Occupant satisfaction in LEED and non-LEED certified buildings. *Building and Environment, 68*, 66–76.

Ametepey, O., Aigbavboa, C., & Ansah, K. (2015). Barriers to successful implementation of sustainable construction in the Ghanaian construction industry. *Procedia Manufacturing, 3*, 1682–1689.

Ariffin, N. A. M., & Gad, S. A. (2015). Biomimicry principles in green building index Malaysia. In *International Joint Conference of SENVAR-iNTA-AVAN, 24–26 November 2015, Johor, Malaysia*. https://core.ac.uk/outputs/300432621

Awadh, O. (2017). Sustainability and green building rating systems: LEED, BREEAM, GSAS and Estidama critical analysis. *Journal of Building Engineering, 11*, 25–29.

Benyus, J. M. (1997). *Biomimicry: Innovation inspired by nature* (Adobe Digital ed.). Australia: HarperCollins.

Berkebile, B., & McLennan, J. (2004). The living building: Biomimicry in architecture, integrating technology with nature. *BioInspire Magazine, 18*, 1–8.

Castro-Lacouture, D., Sefair, J. A., Flórez, L., & Medaglia, A. L. (2009). Optimization model for the selection of materials using a LEED-based green building rating system in Colombia. *Building and Environment, 44*(6), 1162–1170.

Cohen, C., Pearlmutter, D., & Schwartz, M. (2017). A game theory-based assessment of the implementation of green building in Israel. *Building and Environment, 125*, 122–128.

Construction Specifications Institute (2013). *The CSI sustainable design and construction practice guide*. Somerset: John Wiley & Sons.

Doan, D. T., Ghaffarianhoseini, A., Naismith, N., Zhang, T., Ghaffarianhoseini, A., & Tookey, J. (2017). A critical comparison of green building rating systems. *Building and Environment, 123*, 243–260.

Dodo, Y. A., Nafida, R., Zakari, A., Elnafaty, A. S., Nyakuma, B. B., & Bashir, F. M. (2015). Attaining points for certification of green building through choice of paint. *Chemical Engineering, 45*, 1879–1884.

Fuerst, F., & McAllister, P. (2011). Eco-labeling in commercial office markets: Do LEED and Energy Star offices obtain multiple premiums? *Ecological Economics, 70*(6), 1220–1230.

Goh, C. S. (2014). *Development of a capability maturity model for sustainable construction*. The University of Hong Kong.

Goss, J. (2009). *Biomimicry: Looking to nature for design solutions*. Corcoran College of Art Design.

Haapio, A., & Viitaniemi, P. (2008). A critical review of building environmental assessment tools. *Environmental Impact Assessment Review, 28*(7), 469–482.

Hankinson, M., & Breytenbach, A. (2012). Barriers that impact on the implementation of sustainable design. In *The Proceedings of Cumulus Conference, 24–26 May 2012, Helsinki, Finland*.

Iouguina, A. (2013). *Biologically informed disciplines: a comparative analysis of terminology within the fields of bionics, biomimetics, and biomimicry*. Doctoral dissertation, Carleton University.

Jawali, R., & Fernández-Solís, J. L. (2008). A building sustainability rating index (BSRI) for building construction. In *Proceedings of the 8th International Post Graduate Research Conference*, 1–16.

Kibert, C. J. (2013). *Sustainable construction: Green building design and delivery*. John Wiley & Sons.

Kim, M. J., Oh, M. W., & Kim, J. T. (2013). A method for evaluating the performance of green buildings with a focus on user experience. *Energy and Buildings, 66*, 203–210.

Krajangsri, T., & Pongpeng, J. (2018). A comparison of green building assessment systems. *MATEC Web of Conferences, 192*, 1–4.

Kubba, S. (2016). *LEED v4 practices, certification, and accreditation handbook* (2nd edition. ed.). US: Butterworth-Heinemann.

Larsson, N. (2004). An overview of green building rating and labelling systems. *Paper presented at the Symposium on Green Building Labelling*, 15–21.

Lee, W. L. (2013). A comprehensive review of metrics of building environmental assessment schemes. *Energy and Buildings, 62*, 403–413.

Li, B., Li, Y., Yu, W., & Yao, R. (2016). A multidimensional model for green building assessment: A case study of a highest-rated project in Chongqing. *Energy & Buildings, 125*, 231–243.

Li, Y., Chen, X., Wang, X., Xu, Y., & Chen, P. (2017). A review of studies on green building assessment methods by comparative analysis. *Energy and Buildings, 146*, 152–159.

Liu, T. Y., Chen, P. H., & Chou, N. N. (2019). Comparison of assessment systems for green building and green civil infrastructure. *Sustainability, 11*(7), 2117.

Madurwar, M. V., Ralegaonkar, R. V., & Mandavgane, S. A. (2013). Application of agro-waste for sustainable construction materials: A review. *Construction and Building Materials, 38*, 872–878.

Molinaro, R., Corbo, C., Martinez, J. O., Taraballi, F., Evangelopoulos, M., Minardi, S., & Tasciotti, E. (2016). Biomimetic proteolipid vesicles for targeting inflamed tissues. *Nature Materials, 15*(9), 1037–1046.

Nychka, J. A., & Chen, P. (2012). Nature as inspiration in materials science and engineering. *JOM Journal of the Minerals, Metals and Materials Society, 64*(4), 446–448.

Poveda, C. A., & Lipsett, M. (2011). A review of sustainability assessment and sustainability/environmental rating systems and credit weighting tools. *Journal of Sustainable Development, 4*(6), 36.

Said, F. S., & Harputlugil, T. (2019). A research on selecting the green building certification system suitable for Turkey. *GRID-Mimarlık, Planlama Ve Tasarım Dergisi, 2*(1), 25–53.

Serpell, A., Kort, J., & Vera, S. (2013). Awareness, actions, drivers and barriers of sustainable construction in Chile. *Technological and Economic Development of Economy, 19*(2), 272–288.

Siebers, R., Kleist, T., Lakenbrink, S., Bloech, H., den Hollander, J., & Kreißig, J. (2016). *Sustainability certification labels for buildings*. UK: Wiley Online Library.

Suzer, O. (2015). A comparative review of environmental concern prioritization: LEED vs other major certification systems. *Journal of Environmental Management, 154*, 266–283.

Thilakaratne, R., & Lew, V. (2011). Is LEED leading Asia? An analysis of global adaptation and trends. *Procedia Engineering, 21*, 1136–1144.

Wu, P., Feng, Y., Pienaar, J., & Xia, B. (2015). A review of benchmarking in carbon labelling schemes for building materials. *Journal of Cleaner Production, 109*, 108–117.

Yılmaz, M., & Bakış, A. (2015). Sustainability in construction sector. *Procedia-Social and Behavioral Sciences, 195*, 2253–2262.

Zuo, J., & Zhao, Z. (2014). Green building research – current status and future agenda: A review. *Renewable and Sustainable Energy Reviews, 30*, 271–281.

Part II

Sustainability Trends in Construction

2 Sustainability Nexus in the Construction Industry

Introduction

Buildings and other infrastructure have a significant impact on the natural and human environment (health, living conditions, and social well-being of the people). As the provider of this infrastructure, the construction industry (CI) has become a major sector and driver of economic and social developments in both developing and developed countries. In other to stabilise the economy, infrastructural investments by the government authenticate the industry's indispensable role in the national development plan (Giang & Sui Pheng, 2011). Construction and the operation of homes and offices account for a staggering one-tenth of the global economy, according to World Watch (Hill et al., 2002). Also, for example, the CI has accounted for 7% of China's gross domestic product (GDP) since the mid-1990s, according to the China Statistics Press in 2000 and 2001 (Zou et al., 2007). However, with so much focus on quality, time, and cost, the CI has neglected its impact on the environment (Azis et al., 2012). With the continuous industrialisation and urbanisation in most countries of the world, the devastating effect of construction activities on the natural and human environment is on the rise (Peng, 2019). Every building project has an increasingly consequential effect on the natural environment. By significantly contributing to landfill waste, energy consumption, and loss of biodiversity and natural resources, the CI is now perceived in a negative light (Moussa, 2019).

State of the Construction Sector

According to Hamid et al. (2008), the CI has been perceived in the three-dimensional view as demeaning, dangerous, and dirty. Globally, this perception is because of the various adverse impacts of the industry on both the human and natural environment. One of the negative impacts of the industry is waste generation. Waste in the CI poses a serious threat to the productivity, quality, cost, and completion time of projects (Azis et al., 2012). With few exceptions, most of the developed countries employ traditional waste management systems which rely on landfills and consequently pollute the environment in a significant way (Ali & Kumar, 2019). The production of traditional construction materials such as bricks, steel, and cement

DOI: 10.1201/9781003415961-4

(Madurwar et al., 2013) is another avenue through which the CI generates waste. In Asia, 30% of municipal solid waste comes from the CI owing to rapid urbanisation, economic development, and high population density (Li, 2013). Recycling wastes generated from construction materials and activities is one of the practical solutions to waste pollution (Raut et al., 2011) is therefore impossible since these materials are unsustainable in the first place.

The emission of carbon dioxide (CO_2), greenhouse gases (GHGs), and other volatile organic compounds (VOCs) into the atmospheric space is another major issue attributed to the activities of the CI. According to Carmichael et al. (2018), the method of construction employed to undertake a project will determine the quantity of emissions that will be generated. The CI is acknowledged as the largest source of carbon emissions through buildings and construction activities (Wu et al., 2014) without consideration for the environment. It has been largely agreed that these emissions contribute to global warming and climate change. The production of ordinary portland cement (OPC) which is the most widely used construction material contributes about 5% of man-made CO_2 emissions globally (Damtoft et al., 2008; Sivakrishna et al., 2019). From 50.4 million tons of CO_2 in 2013 to 51.7 million tons in 2017, emissions in Austria have increased by 3.3% which is due to increased reliance on fossil fuels and other construction activities (Fadai & Stephan, 2019). Also, CO_2 emissions released from the CI in China are reported to be a 6.57% increase per year over the past decade (Ma & Cai, 2019), which is largely due to the growing demand for household energy services (Ma et al., 2019).

Excessive resource consumption and depletion of natural resources due to a growing demand for raw materials is another attribute ascribed to the CI. The rapid population growth experienced in most developing countries keeps exerting a huge demand on the natural environment (Ekong, 2017). Triggered by the increasing demand for infrastructural facilities, the CI is changing land reforms, leading to the depletion of natural resources happening at a faster and more alarming rate (Lam et al., 2010). Plant and animal life as well as aquatic systems are being disrupted and constantly threatened as the need to replace them with buildings, roads, and other basic amenities is on the rise. Land use requirements in the CI are claimed to change the biodiversity and compete with arable land for agricultural purposes (Evans et al., 2009). It is reported that about 40% of the raw materials extracted globally can be traced to the CI with around three billion tons of raw materials utilised in the manufacture of construction materials on an annual basis (Baglou et al., 2017). In China, construction activities due to the high rate of urbanisation have increased the quantity of annual land development by consuming more than half of the country's natural resources, thereby destroying the ecological environment (Hua & Min, 2019).

Excessive energy consumption and water usage are other adverse environmental impacts of the CI. While energy use is important to human well-being and results in substantial economic growth, the environmental externalities (emissions of pollutants) throughout its lifecycle are devastating (Liao et al., 2019). About 3% of the total energy consumed globally is traceable to concrete and other construction materials utilised throughout the life of a building (Damtoft et al., 2008).

For instance, the sum of energy consumed by buildings in China is linked to the country's promotion of large-scale urbanisation and the construction of new rural areas (Zhong, 2019). As reported by the United Nations Environment Programme (UNEP), CI globally accounts for up to 40% of the total energy consumed (Hwang et al., 2016). Reports from Kuwait likewise indicated that construction consumes significant amounts of water, including that desalinated in power plants (AlSanad, 2015). The study by Hong et al. (2019) indicated that 27.20% of embodied energy consumed and 8.97% of virtual water usage in China are accounted for by the CI.

The high rate of accidents and other work-related injuries, particularly musculoskeletal disorders (MSDs), remains rampant in the CI (Glimskär & Lundberg, 2013). Despite being criticised and regarded as a low-quality, less productive (Yusof et al., 2014), and hazardous sector, the industry is continually plagued with frequent occurrences of accidents and other injuries. From small-scale to large-scale accidents with a high frequency, the CI is known to be a source of diverse hazards (Zhou et al., 2015). As reported in the study by Zhang et al. (2019), a total of 3821 construction accidents occurred in China between 2010 and 2016. In Hong Kong, a total of 3723 accidents and 62% of industrial fatalities reported are attributed to the CI (Yiu et al., 2019). In Australia, construction accidents are not only prevalent but also have significant financial implications due to compensations, legal penalties, and medical payments, among others (Allison et al., 2019). With a significant impact on construction workers, construction firms, and the quality and success of construction projects, accidents and work-related injuries in the CI remain one of the adverse impacts of the industry.

Other issues raised regarding the CI include deforestation, a high rate of corruption, inequality, poor and unfavourable work conditions, poor wages and benefits, non-adherence to best construction practices and standards, exploitation of the construction workforce, and a disregard for health and safety (H&S) procedures, among others. As suggested by Wever and Vogtländer (2015), a better method of process, production, and consumption is imperative to disengage societal progress from environmental disintegration and to embrace the utilisation of non-renewable resources. Hence, the introduction of the concept of sustainability or sustainable development (SD) aims to address the ills and other adverse impacts of the CI by encouraging efficient processes, responsible use of resources, and consideration for the future.

Understanding Sustainability and Sustainable Development

In the book titled 'A New Environmental Ethics: The Next Millennium for Life on Earth', Prof. Holmes Rolston III made a significant observation about the unforeseen magnitude and importance of the environmental crisis that unfolded in the latter half of the previous century. He emphasised that as we entered the current century, the environmental crisis became impossible to ignore (Rolston, 2012). This crisis encompasses critical challenges such as indoor environmental quality issues, climate change, global warming, loss of biodiversity, water shortages, and pollution of land, water, and air, among many others. It is in response to these challenges that the concepts of sustainability and SD have emerged. The concept of SD is considered an alternative and response to the shortcomings of conventional development, which has failed to distribute wealth fairly and has had adverse impacts on the environment, social cohesion, and cultural diversity (Villeneuve et al., 2017).

Although the term 'green' is often used interchangeably with sustainability, various sources of literature assert that sustainability or SD encompasses a comprehensive interconnectedness and considers social, economic, and environmental considerations, while 'green' primarily focuses on environmental aspects (Diófási & Valkó, 2011; Agbedahin, 2019; Khoshnava et al., 2019). In a study on housing development conducted by Turcotte (2006), the idea that 'green' is synonymous with 'sustainable' is critiqued, and it is concluded that sustainability or SD goes beyond mere environmental concerns. Waas et al. (2011) further affirm that the terms 'sustainability' and 'sustainable development' have been misused and misunderstood, particularly with 'green', resulting in diminished effectiveness in implementing the core principles by governments, individuals, and organisations. While achieving environmental goals such as resource conservation, energy, and water efficiency, the use of less toxic materials and waste reduction, and minimisation are vital steps towards realising SD, it is important to recognise that other factors, particularly social and economic considerations, are equally imperative. Therefore, in this book chapter, the terms 'sustainability' and 'sustainable development' are adopted, acknowledging their holistic nature and their inclusion of social, economic, and environmental dimensions.

The concept of SD gained prominence in the late 20th century, although its roots can be traced back much further. It gained significant attention in 1987 with the release of the influential report titled Our Common Future by the World Commission on Environment and Development (WCED), also known as the Brundtland Commission (Suliman & Omran, 2009; Scarlett, 2010; Waris et al., 2019). Since then, SD has evolved into a globally accepted and recognised concept across all sectors and fields of human endeavour. As emphasised by Waas et al. (2011) and the study conducted by Weerasinghe (2012), several key milestones have contributed to the advancement of SD. These include the United Nations Conference on the Human Environment (UNCHE) in 1972, the World Conservation Strategy (WCS) in 1980, the WCED in 1987, the United Nations Conference on Environment and Development (UNCED) in 1992, the United Nations Millennium Summit and the Earth Charter in 2000, the United Nations World Summit on Sustainable Development (WSSD) in 2002, and the Rio+20 United Nations Conference on Sustainable Development (UNCSD) in 2012. These global and political initiatives have demonstrated concerted efforts towards the implementation of SD.

To gain a comprehensive understanding of the fundamentals and objectives of SD, numerous scholars have put forward different definitions, making it challenging to align with a specific one. Each of these definitions portrays SD as a flexible concept that reflects the diverse perspectives of scholars regarding the global challenges faced by humanity (Krueger, 2017). However, among all the definitions, the definition provided by the Brundtland Commission is considered the most suitable and widely cited. The Brundtland Commission defines SD as 'development that meets the needs of the present without compromising the ability of future genera- tions to meet their own needs' (WCED, 1987; Licon, 2004; Weerasinghe, 2012; Opon & Henry, 2019; Topak et al., 2019). Other definitions of SD are often derived from the Brundtland Commission's definition with minor alterations and slight modifications (Roostaie et al., 2019). Hendiani and Bagherpour (2019), as well as Roca-Puig (2019), define SD as the ability to sustain economic growth, environ- mental protection, and social inclusion in a human system over time. The defini- tion proposed by Geissdoerfer et al. (2016) and Pieroni et al. (2019) describes SD as the integration of environmental resilience, social inclusiveness, and economic performance with equity, for the benefit of present and future generations. SD can be understood as adjusting human actions to address the needs of the present gen- eration while safeguarding the ability of future generations to address their own needs (Opoku & Ahmed, 2014). A careful examination of these various definitions reveals a common trend and demand for balanced attention to the three pillars of sustainability: social, environmental, and economic dimensions.

Models of Sustainability and Sustainable Development

Sustainability and SD encompass a trio of core elements: economic growth, social well-being, and environmental conservation and preservation (Ngan et al., 2019). In a Schumacher Lecture delivered by Christopher Alexander (2004), he grouped the key issues of SD into technical and philosophical aspects. Another perspec- tive on sustainability is presented by Wever and Vogtländer (2015) through the Tripe-P model, which identifies the importance of people (social aspects), planet (ecological consequences), and profit (economic viability). Holden et al. (2017) perceive SD in the triple dimension of equity (ensuring social equity), needs (meet- ing human needs), and limits (respecting environmental boundaries). However, the United Nations (UN) 2030 Agenda for SD, known as the Sustainable Development Goals (SDGs), takes a 5-Ps approach, including people, planet, prosperity, peace, and partnership. People, planet, and prosperity reflect the three dimensions of SD, while peace and partnership serve as essential conditions and pathways for achiev- ing the SDGs (United Nations, 2015). These different frameworks and perspec- tives highlight the multi-faceted nature of sustainability and SD, encompassing economic, social, and environmental considerations. Whether through the 3-Ps, 5-Ps, or other conceptual models, the aim remains to foster a harmonious balance between human well-being, ecological integrity, and long-term economic viability. In conclusion, existing literature widely agrees that the trio of economic, societal, and environmental factors are crucial considerations in achieving SD (Iwaro & Mwasha, 2013). dimensions.

Social Dimension of Sustainability and Sustainable Development

Social sustainability is concerned with maintaining the overall viability and smooth functioning of social systems (Munasinghe, 2004). Often overlooked, societal sustainability, which encompasses the social dimension of SD, plays a critical role in establishing a sustainable environment (Wickizer & Snow, 2010). This pillar focuses on enhancing both individual and collective well-being within society by promoting social capital, which refers to the capacity of individuals and communities to collaborate effectively. The objective is to foster the development of just and equitable societies, where positive human growth is nurtured, providing opportunities for a satisfactory quality of life and self-fulfilment (Du Plessis, 2002). According to Othman (2010), the social dimension of sustainability encompasses various aspects such as workers' health and safety, benefits for marginalised groups (e.g., disabled individuals and low-income earners), adherence to international and national laws, global poverty alleviation and citizen action, urban planning, and transportation, the relationship between human rights and human development, impacts on local communities and quality of life, corporate power and environmental justice, and local and individual lifestyles, and ethical consumerism. Whang and Kim (2015) identify key factors to consider within the social element of sustainability, including employment, health and safety, well-being, education/training, partnership working, culture/heritage, security, community, and service quality. According to Alexander (2004), the following core issues should be considered within the social and philosophical elements of sustainability:

- Implementing birth control measures to help reduce and stabilise the Earth's population.
- Preserving endangered and threatened species from extinction.
- Safeguarding the natural ecology by protecting plant and animal life in their relationship to human life.
- Cultivating a spiritually healthy relationship between inhabitants, users, communities, and their environment.
- Introducing the economics of sustainable thinking to mitigate the negative impacts of large-scale corporate development.
- Promoting the physical and social health of the environment.

These considerations highlight the importance of addressing social and philosophical aspects in achieving SD. They emphasise the well-being of both humans and the natural environment, recognising the interconnectedness and interdependence between them.

Economic Dimension of Sustainability and Sustainable Development

The economic element of sustainability entails the development of a system that promotes equal access to opportunities and resources, along with the fair distribution of limited eco-friendly productive space based on ethical principles (Du Plessis, 2002). As described by Munasinghe (2004), economic sustainability involves maximising income generation while maintaining the stock of assets (capital) that yield beneficial outputs. It focuses on creating material well-being and financial stability, ultimately aiming to enhance the capacity to provide goods and services that satisfy human needs and contribute to an improved standard of living (Iwaro & Mwasha, 2013). The economic dimension of sustainability is reflected in Goals 1 (ending poverty in all its forms everywhere) and Goals 8 (promoting sustained, inclusive, and sustainable economic growth, full and productive employment, and decent work for all) of the United Nations' SDGs (United Nations, 2015).

According to Othman (2010), the economic dimension of sustainability includes improving the quality of life, creating new markets and sales opportunities, integrating ecological concerns with social and economic considerations, supporting local businesses, generating additional value, reducing costs through increased efficiency and lower energy and raw material inputs, and enhancing market share through an improved public image. Whang and Kim's study (2015) identified several important factors to consider within the economic element of sustainability, such as competitiveness, productivity/profitability, partnerships, project delivery, value for money, knowledge management, retention of skilled labour, life cycle cost, construction cost, affordability, support of the local economy, quality management for durability, commercial viability, innovation/research and development (R&D), and image and reputation. To achieve economic profitability within the context of SD, which aims to enhance the quality of life for humanity, Suliman and Omran (2009) as well as Zabihi and Habib (2012) emphasised the importance of addressing the following objectives comprehensively:

- Maintenance of stable levels of employment and economic growth to the maximum extent possible.
- Effective protection of the natural environment.
- Promotion of social progress that recognises the needs of people.
- Efficient utilisation of natural resources.

These considerations underline the significance of economic sustainability in achieving a balanced and thriving society, where resources are used efficiently, wealth is distributed equitably, and economic activities support long-term well-being and growth. By integrating these objectives holistically, SD strives to strike a balance between economic prosperity, social well-being, and environmental conservation. This approach ensures the efficient utilisation of resources, considers the needs and aspirations of individuals and communities, preserves the integrity of the natural environment, and fosters sustainable economic growth while maintaining stable employment levels.

Environmental Dimension of Sustainability and Sustainable Development

A few examples of the numerous environmental issues that pose a threat to our planet include deforestation, pollution, emissions (such as GHGs and VOCs), depletion and degradation of natural resources, and loss of biodiversity. These issues render natural organisms vulnerable and less resilient, hindering their ability to regenerate and survive. Environmental sustainability is centred around preserving the overall viability of natural systems, improving their health, and enhancing their capacity to adapt to changes across different timeframes and geographical scales (Munasinghe, 2004). This dimension of sustainability involves various actions, including the use of renewable raw materials, reducing impacts on human health, eliminating toxic substances, minimising waste generation and emissions, and preserving ecosystems (Othman, 2010). It emphasises the importance of energy conservation, resource efficiency, water efficiency, responsible land use, effective waste management, maintaining atmospheric quality, promoting indoor environmental quality, sustainable transportation, preserving ecological environments, and effective environmental management (Whang & Kim, 2015).

According to Alexander (2004), the following core issues should be considered within the environmental element of sustainability:

• Establishing renewable and non-destructive cycles of food production, material production, and land management.
• Implementing water and waste management practices that prioritise water recycling and utilise refuse and waste for land fertilisation.
• Developing non-destructive energy sources such as solar energy, tidal energy, and wind energy.
• Preserving and restoring bioregions.
• Protecting and promoting the recycling of natural resources.
• Taking measures to protect planetary climate stability.
• Safeguarding soil and water resources from exploitation and erosion.
• Reducing wasteful energy consumption.
• Utilising appropriate green building (GB) materials.

These considerations emphasise the importance of adopting sustainable practices to protect the environment and ensure the responsible use of natural resources. By implementing measures such as the 3Rs (reduce, reuse, and recycle), using environmentally friendly materials, minimising energy waste, protecting soil and water resources, addressing climate change, promoting recycling, preserving ecosystems, and utilising renewable energy sources, the environmental dimension of sustainability aims to minimise the negative impacts of human activities on the planet and foster a harmonious relationship between human development and the natural environment.

Overview of the United Nations Sustainable Development Goals

Achieving SD remains an ambitious agenda that strives to include everyone in the process of its realisation. Given its significance in addressing the challenges facing humanity today, a wide array of stakeholders including scientists, researchers, agencies, civil society organisations, corporate and private sectors, as well as multilateral and bilateral development partners, are actively shaping the global perception of SD. As highlighted by Agbedahin (2019), it is crucial to make tough and rigorous decisions, plans, and strategies at international, regional, national, and local levels, supported by good governance and political will, to effectively implement SD. As the latest global effort aimed at achieving SD, the SDGs present a streamlined and comprehensive 2030 Agenda initiated by the United Nations (UN). All countries and stakeholders operating under the auspices of the UN are expected to adopt and implement this plan through collaborative partnerships. Outlined in the report titled 'Transforming Our World', the SDGs were adopted by the General Assembly of UN member states in 2015 with the aim of catalysing actions for SD over the next 15 years, focusing on vital areas for the planet and humanity (United Nations, 2015; Holden et al., 2017).

Figure 2.1 provides a summarised list of the SDGs and their respective objectives. The SDGs comprise 17 goals and 169 targets, reflecting the ambitious and comprehensive nature of the 2030 Agenda for Sustainable Development. These goals and targets emerged from years of collaborative efforts by countries, the United Nations (UN), and the lessons learned from the previous Millennium Development Goals (MDGs), which highlighted both successes and shortcomings in driving global progress towards resilience and sustainability. However, some scientists and researchers argue that the messages conveyed by the SDGs can be ambiguous, with the constituents – 17 goals, 169 targets, and 330 indicators – sometimes perceived as vague (Dalampira & Nastis, 2020). Since the primary focus of SD is human welfare (the 'people' dimension), the SDGs must be understandable to the people, who are the key stakeholders required to drive the agenda. Furthermore, each country and region should be allowed to develop a customised plan for achieving the SDGs and targets, aligning them with global objectives. It is important to consider the unique national or regional policies, the socio-cultural status of the citizens, circumstances, developmental priorities, and capacities when pursuing the SDGs and their targets. The study conducted by Olawumi and Chan (2018) also supports the notion of developing country-specific SD policies while maintaining a global perspective. The successful implementation of the SDGs should therefore be driven by the collective effort of global entities with local agencies, government, and indigenous people in achieving SD, through collaboration, shared responsibility, and long-term planning for the betterment of our planet and future generations.

Summary

This chapter presents an overview of the state of the CI globally looking at the various adverse impacts it has on the human and natural environments. The chapter also presents the concept of sustainability and SD to understand its fundamentals. Also, the chapter examines the models of sustainability by providing the tenets and objectives of the social, economic, and environmental dimensions of sustainability and SD. The next chapter presents the concept of sustainability in the construction sector by examining the various trends and practices for optimising sustainability in the sector.

Figure 2.1 Summarised list and objectives of the SDGs

References

Agbedahin, A. V. (2019). Sustainable development, education for sustainable development, and the 2030 Agenda for Sustainable Development: Emergence, efficacy, eminence, and future. *Sustainable Development, 27*(4), 669–680.

Alexander, C. (2004). *Sustainability and morphogenesis: The birth of a living world.* Bristol: Centre for Environmental Structure.

Ali, S. R., & Kumar, R. (2019). Strategic framework and phenomenon of zero waste for sustainable future. *Contaminants in Agriculture and Environment: Health Risks and Remediation, 1*, 200.

Allison, R., Hon, C. K. H., & Xia, B. (2019). Construction accidents in Australia: Evaluating the true costs. *Safety Science, 120*, 886–896.

AlSanad, S. (2015). Awareness, drivers, actions, and barriers of sustainable construction in Kuwait. *Procedia Engineering, 118*, 969–983.

Azis, A. A. A., Memon, A. H., Rahman, I. A., Nagapan, S., & Latif, Q. B. A. I. (2012). Challenges faced by construction industry in accomplishing sustainablity goals. In *The Proceedings of 2012 IEEE Symposium on Business, Engineering and Industrial Applications, 23–26 September 2012, Bandung, Indonesia* (pp. 630–634). IEEE.

Baglou, M., Ghoddousi, P., & Saeedi, M. (2017). Evaluation of building materials based on sustainable development indicators. *Journal of Sustainable Development, 10*(4), 143–154.

Carmichael, D. G., Mustaffa, N. K., & Shen, X. (2018). A utility measure of attitudes to lower-emissions production in construction. *Journal of Cleaner Production, 202*, 23–32.

Dalampira, E. S., & Nastis, S. A. (2020). Mapping sustainable development goals: A network analysis framework. *Sustainable Development, 28*(1), 46–55.

Damtoft, J. S., Lukasik, J., Herfort, D., Sorrentino, D., & Gartner, E. M. (2008). Sustainable development and climate change initiatives. *Cement and Concrete Research, 38*(2), 115–127.

Diófási, O., & Valkó, L. (2011). Green (public) procurement in practice-methods and tools for the successful implementation. *Regional and Business Studies, 3*(2 Suppl), 11–23.

Du Plessis, C. (2002). *Agenda 21 for sustainable construction in developing countries* (No. 204). Pretoria, South Africa: CSIR Building and Construction Technology.

Ekong, F. U. (2017). Applying the concept of green cities in Nigeria: Challenges and prospects. *Advances in Social Sciences Research Journal, 4*(10), 85–96.

Evans, A., Evans, T. J., & Strezov, V. (2009). Assessment of sustainability indicators for renewable energy technologies. *Renewable and Sustainable Energy Reviews, 13*(5), 1082–1088.

Fadai, A., & Stephan, D. (2019). Ecological performance and recycling options of primary structures. *IOP Conference Series: Earth and Environmental Science, 323*(1), 1–9.

Geissdoerfer, M., Bocken, N. M., & Hultink, E. J. (2016). Design thinking to enhance the sustainable business modelling process – a workshop based on a value mapping process. *Journal of Cleaner Production, 135*, 1218–1232.

Giang, D. T. H., & Sui Pheng, L. (2011). Role of construction in economic development: Review of key concepts in the past 40 years. *Habitat International, 35*(1), 118–125.

Glimskär, B., & Lundberg, S. (2013). Barriers to adoption of ergonomic innovations in the construction industry. *Ergonomics in Design, 21*(4), 26–30.

Hamid, A. R. A., Majid, M. Z. A., & Singh, B. (2008). Causes of accidents at construction sites. *Malaysian Journal of Civil Engineering, 20*(2), 242–259.

Hendiani, S., & Bagherpour, M. (2019). Developing an integrated index to assess social sustainability in construction industry using fuzzy logic. *Journal of Cleaner Production, 230*, 647–662.

Hill, R., Bowen, P., & Opperman, L. (2002). Sustainable building assessment methods in South Africa: An agenda for research. *Sustainable building* (pp. 1–6). Rotterdam (Netherlands): In-house publishing.

Holden, E., Linnerud, K., & Banister, D. (2017). The imperatives of sustainable development. *Sustainable Development, 25*, 213–226.

Hong, J., Zhong, X., Guo, S., Liu, G., Shen, G. Q., & Yu, T. (2019). Water-energy nexus and its efficiency in China's construction industry: Evidence from province-level data. *Sustainable Cities and Society, 48*, 1–11.

Hua, L., & Min, Z. (2019). Carbon emission efficiency of construction industry in Hunan province and measures of carbon emission reduction. *Nature Environment & Pollution Technology, 18*(3), 1005–1010.

Hwang, B., Zhu, L., & Ming, J. T. T. (2016). Factors affecting productivity in green building construction projects: The case of Singapore. *Journal of Management in Engineering, 33*(3), 1–12.

Iwaro, J., & Mwasha, A. (2013). The impact of sustainable building envelope design on building sustainability using integrated performance model. *International Journal of Sustainable Built Environment, 2*(2), 153–171.

Khoshnava, S. M., Rostami, R., Zin, R. M., Štreimikienė, D., Yousefpour, A., Strielkowski, W., & Mardani, A. (2019). Aligning the criteria of green economy (GE) and sustainable development goals (SDGs) to implement sustainable development. *Sustainability, 11*(17), 4615.

Krueger, R. (2017). Sustainable development. In Richardson, D., N. Castree, M. F. Goodchild, A. Kobayashi, W. Liu, & R. A. Marston (Eds.), *International encyclopedia of geography* (pp. 1–14). John Wiley & Sons.

Lam, P. T. I., Chan, E. H. W., Poon, C. S., Chau, C. K., & Chun, K. P. (2010). Factors affecting the implementation of green specifications in construction. *Journal of Environmental Management, 91*(3), 654–661.

Li, Y. (2013). *Developing a sustainable construction waste estimation and management system.* Hong Kong University of Science and Technology.

Liao, X., Chai, L., Ji, J., Mi, Z., Guan, D., & Zhao, X. (2019). Life-cycle water uses for energy consumption of Chinese households from 2002 to 2015. *Journal of Environmental Management, 231*, 989–995.

Licon, C. V. (2004). *An evaluation model of sustainable development possibilities.* Arizona State University.

Ma, M., & Cai, W. (2019). CO2 mitigation model for China's residential building sector. In *The Proceedings of Applied Energy Symposium, 16–18 October 2019, Xiamen, China.*

Ma, M., Ma, X., Cai, W., & Cai, W. (2019). Carbon-dioxide mitigation in the residential building sector: A household scale-based assessment. *Energy Conversion and Management, 198*(1–15), 111915.

Madurwar, M. V., Ralegaonkar, R. V., & Mandavgane, S. A. (2013). Application of agro-waste for sustainable construction materials: A review. *Construction and Building Materials, 38*, 872–878.

Moussa, R. R. (2019). The reasons for not implementing green pyramid rating system in Egyptian buildings. *Ain Shams Engineering Journal, 10*(4), 917–927.

Munasinghe, M. (2004). Sustainable development: Basic concepts and application to energy. *Encyclopedia of Energy, 5*, 789–808.

Ngan, S. L., How, B. S., Teng, S. Y., Promentilla, M. A. B., Yatim, P., Er, A. C., & Lam, H. L. (2019). Prioritization of sustainability indicators for promoting the circular economy: The case of developing countries. *Renewable and Sustainable Energy Reviews, 111*, 314–331.

Olawumi, T. O., & Chan, D. W. (2018). A scientometric review of global research on sustainability and sustainable development. *Journal of Cleaner Production, 183*, 231–250.

Opoku, A., & Ahmed, V. (2014). Embracing sustainability practices in UK construction organizations: Challenges facing intra-organizational leadership. *Built Environment Project and Asset Management, 4*(1), 90–107.

Opon, J., & Henry, M. (2019). An indicator framework for quantifying the sustainability of concrete materials from the perspectives of global sustainable development. *Journal of Cleaner Production, 218*, 718–737.

Othman, A. A. E. (2010). Incorporating innovation and sustainability for achieving competitive advantage in construction. In Wallis, I., L. Bilan, M. Smith & A. S. Kazi (Eds.), *Industrialised, integrated, intelligent sustainable construction I3CON handbook 2*, 13–42, I3CON in collaboration with BSRIA.

Peng, J. J. (2019). Factors affecting carbon emissions in the construction industry based on STIRPAT model: Taking Henan province of China as an example. *Nature Environment & Pollution Technology, 18*(3), 1035–1040.

Pieroni, M. P., McAloone, T., & Pigosso, D. A. (2019). Business model innovation for circular economy and sustainability: A review of approaches. *Journal of Cleaner Production, 215*, 198–216.

Raut, S. P., Ralegaonkar, R. V., & Mandavgane, S. A. (2011). Development of sustainable construction material using industrial and agricultural solid waste: A review of waste-create bricks. *Construction and Building Materials, 25*(10), 4037–4042.

Roca-Puig, V. (2019). The circular path of social sustainability: An empirical analysis. *Journal of Cleaner Production, 212*, 916–924.

Rolston, H. III (2012). *A new environmental ethics: The next millennium for life on earth*. UK: Routledge.

Roostaie, S., Nawari, N., & Kibert, C. J. (2019). Sustainability and resilience: A review of definitions, relationships, and their integration into a combined building assessment framework. *Building and Environment, 154*, 132–144.

Scarlett, L. (2010). Cities and sustainability-ecology, economy, and community. *Sustainable Development Law & Policy, XI*(1), 1–74.

Sivakrishna, A., Adesina, A., Awoyera, P. O., & Kumar, K. R. (2019). Green concrete: A review of recent developments. *Materials Today: Proceedings, 27*, 54–58.

Suliman, L. K. M., & Omran, A. (2009). Sustainable development and construction industry in Malaysia. *University of Bucharest, Faculty of Business & Administration, 1*(10), 76–85.

Topak, F., Tokdemir, O. B., Pekericli, M. K., & Tanyer, A. M. (2019). Sustainable construction in Turkish higher education context. *Journal of Construction Engineering, Management & Innovation, 2*(1), 40–47.

Turcotte, D. A. (2006). *A framework for sustainable housing development in the United States*. University of Massachusetts Lowell.

United Nations. (2015). *Transforming our world: The 2030 Agenda for Sustainable Development*. Retrieved 24 September 2019 from https://sustainabledevelopment.un.org/post2015/transformingourworld

Villeneuve, C., Tremblay, D., Riffon, O., Lanmafankpotin, G., & Bouchard, S. (2017). A systemic tool and process for sustainability assessment. *Sustainability, 9*(10), 1–29.

Waas, T., Hugé, J., Verbruggen, A., & Wright, T. (2011). Sustainable development: A bird's eye view. *Sustainability, 3*(10), 1637–1661.

Waris, M., Panigrahi, S., Mengal, A., Soomro, M. I., Mirjat, N. H., Ullah, M., Azlan, Z. S., & Khan, A. (2019). An application of analytic hierarchy process (AHP) for sustainable procurement of construction equipment: Multicriteria-based decision framework for Malaysia. *Mathematical Problems in Engineering, 2019*, 1–20.

Weerasinghe, U. G. D. (2012). *Development of a framework to assess sustainability of building projects*. University of Calgary.

Wever, R., & Vogtländer, J. (2015). Design for the value of sustainability. In Van den Hoven, J., P. Vermaas & I. van de Poel (Eds.), *Handbook of ethics, values, and technological design: Sources, theory, values and application domains* (e-book ed., pp. 513–549). Dordrecht: Springer.

Whang, S., & Kim, S. (2015). Balanced sustainable implementation in the construction industry: The perspective of Korean contractors. *Energy and Buildings, 96*, 76–85.

Wickizer, B. J., & Snow, A. (2010). Rediscovering the transportation frontier: Improving sustainability in the United States through passenger rail. *Sustainable Development Law & Policy, XI*(1), 14–16.

World Commission on Environment and Development (1987). *Our common future*. Oxford, UK: Oxford University Press.

Wu, P., Low, S. P., Xia, B., & Zuo, J. (2014). Achieving transparency in carbon labelling for construction materials – Lessons from current assessment standards and carbon labels. *Environmental Science and Policy, 44*, 11–25.

Yiu, N. S. N., Chan, D. W. M., Shan, M., & Sze, N. N. (2019). Implementation of safety management system in managing construction projects: Benefits and obstacles. *Safety Science, 117*, 23–32.

Yusof, N., Mustafa Kamal, E., Kong-Seng, L., & Iranmanesh, M. (2014). Are innovations being created or adopted in the construction industry? Exploring innovation in the construction industry. *Sage Open, 4*(3), 1–9.

Zabihi, H., & Habib, F. (2012). Sustainability in building and construction: Revising definitions and concepts. *International Journal of Emerging Sciences, 2*(4), 570.

Zhang, J., Zhang, W., Xu, P., & Chen, N. (2019). Applicability of accident analysis methods to Chinese construction accidents. *Journal of Safety Research, 68*, 187–196.

Zhong, W. X. (2019). Influencing factors of the energy consumption behaviour of civil buildings in Hubei province, China. *Nature Environment & Pollution Technology, 18*(3).

Zhou, Z., Goh, Y. M., & Li, Q. (2015). Overview and analysis of safety management studies in the construction industry. *Safety Science, 72*, 337–350.

Zou, P. X. W., Fang, D., Qing Wang, S., & Loosemore, M. (2007). An overview of the Chinese construction market and construction management practice. *Journal of Technology Management in China, 2*(2), 163–176.

3   Sustainable Construction Trends and Practices

Introduction

The several adverse impacts of the construction industry (CI) on the human and natural environment have made the industry a major sector where the sustainability concept is to be adopted and implemented. Environmental issues such as excessive waste generation, greenhouse gas (GHG) emissions, high consumption of energy and water, pollution of various kinds, and the depletion and degradation of natural resources, among others, are the obvious impacts of the CI; hence the need to embrace measures for mitigation. Sustainable construction (SC) intends to proffer solutions not only to the environmental challenges in the CI but also to the social and economic concerns (Ashour et al., 2015). Despite being perceived as an expensive concept, SC compared to the traditional practices in the CI provides opportunities for competitiveness, innovation, and growth that are environmentally conscious. The clamour to adopt and implement SC has also increased over the years as a means of minimising the environmental effects of rapid population and economic growth resulting in increased pressure on natural resources (Saleh & Alalouch, 2015). To this end, this chapter examines the concept of sustainability as it relates to the CI with a keen interest in the various trends and practices aimed at ticking the social, environmental, and economic boxes of sustainability in the sector.

Sustainability in the Construction Sector

The issue of sustainability in the CI is a broad and hotly debated subject globally with different experts and researchers postulating diverse aligning philosophies. The terms 'green', 'green building', 'environmental-friendly', 'green construction', 'sustainable building', and 'high performance' have been used interchangeably with SC (Shofoluwe, 2011; Akadiri et al., 2012; Wong, 2015; Davies & Davies, 2017). As described by Kibert et al. (2000), SC is the translation and alignment of the CI with sustainable development (SD) by ensuring construction practices adhere to sustainability principles. As further clarified by Presley and Meade (2010), SC not only refers to eco-friendly buildings but also to the processes and actions (waste management, transportation) towards the construction. The aim is to minimise the negative environmental impacts of construction while maximising the

DOI: 10.1201/9781003415961-5

valuable economic and social impacts (Gibberd, 2002; Plank, 2008). Marhani et al. (2013) defined SC as the birthing and responsible management of a healthy built environment premised on ecological principles and resource efficiency.

SC is the construction that contributes to SD in the CI, with its implementation leading to better well-being of present and future generations and fewer environmental impacts (Said et al., 2009). To achieve SC, sustainability principles must be integrated into processes, activities, and practices in the CI with every stakeholder taking responsibility. For example, the Chinese CI requires a harmonious coexistence between people, nature, and architecture to achieve SD (Li & Li, 2019). As opined by Mousa (2015), transforming the present traditional practices in the CI to a sustainable one is believed to hinge largely on promoting the awareness and understanding of SC principles and related concepts among the stakeholders. As affirmed by Gan et al. (2015), the government as a major stakeholder plays a critical role in ensuring the adoption and implementation of SC. Collaboration and coordination among all the stakeholders in the CI are crucial to the achievement and success of SC (Abdul Nifa et al., 2015). Serpell et al. (2013) also stated that the achievement of SC will require the utilisation of different strategies that are country-specific and tailored to the level of development and specific socioeconomic status of CI.

Principles of Sustainability in the Construction Industry

For the adoption, integration, and implementation of SC to be successful, SD must be broken down into specific design and construction criteria (Szydlik, 2014). These criteria are regarded as SC principles, applicable throughout the lifecycle of building projects. Principles of sustainability in the CI are the framework and guidelines that aid the achievement of SC through environmentally responsive construction processes and activities. An example of such a process highlighted by Sameh (2014) is the replacement of traditional construction materials with sustainable ones to minimise and overcome the negative environmental impacts of the CI. The successful achievement of SC relies on the balanced integration of sustainability principles. Based on a comprehensive review of literature, the study of Goh (2014) presented 11 main principles of SC, namely resource/material consumption (resource usage efficiency, recycling and reuse of materials, use of local/regional materials, land use, and water efficiency); environmental impact (life cycle assessment, waste management, toxic elimination, carbon emission, ecosystem, and greenery); quality of comfort (occupational health and safety, indoor environment quality, indoor chemical and pollutant source control, and controllability of systems); energy efficiency (renewable energy, and optimal energy performance); passive design (daylight, thermal comfort, ventilation, spaces flexibility and adaptability, and ecological innovation); life cycle costing (cost effectiveness, and financial return and payback period); life span (service life/durability of building and design, and maintenance and refurbishment); heritage and cultural preservation (heritage conservation, and regional culture preservation); site aspect (connectivity to adjacent neighbourhood, and transport link to local context); functional applicability (practicability and functional applicability); and social and community benefits (stakeholders' satisfaction). A perusal of these principles revealed they represent the three dimensions (social, environmental, and economic) of SD. According to Al Sanad (2015), the application of SC principles is imperative when evaluating the required elements for construction in line with SD. SC principles apply to resources required to create and operate a building throughout its lifecycle (planning to deconstruction). Table 3.1 presents the principles of SC and the corresponding authors.

Sustainable Construction Trends and Practices

SC practices are no longer new but have become acceptable standards of practice for professionals and stakeholders in the CI. SC practices, trends, and principles offer more opportunities than conventional practices by creating solutions that are environmentally friendly and conscious of human life. They represent different efforts and drives in the CI geared towards achieving the three dimensions of sustainability. SC practices aim to reduce the overall environmental impact of the CI throughout its lifecycle (Opoku & Fortune, 2015). Owing to the complexities attached to achieving SD in the CI, some of the efforts (SC practices, trends, and principles) are focused on addressing the whole dimension while the majority concentrate only on the environmental issues. Globally, it is estimated that more than half of the studies and research on SC only addressed environmental issues, leaving a huge knowledge gap on social and economic concerns (Whang & Kim, 2015). While some of these trends and practices have been in existence and utilised over the years, there are now a number of them that emanated with the fourth industrial revolution (4IR) era as technological advancements for achieving sustainability in the built environment.

As listed by Huovila and Koskela (1998), Alves et al. (2012), and Blizzard and Klotz (2012), sustainable principles in the CI include biomimicry principles, lean construction (LC), cradle-to-cradle design principles, sustainable design principles, ecological design principles, whole systems thinking principles, integrated design principles, sustainable engineering principles, whole systems approach, green design principles, SD design principles, design for the environment principles, and green engineering (GE) design principles. Referred to as SC concepts, Kibert and Fenner (2017) highlighted the following: whole-building design, biomimicry, SC, passive design, net-zero energy buildings (NZEBs), biophilia; life-cycle assessment; resilience, environmental product declarations (EPDs), ecological footprint, ecological rucksack, and embodied energy. Sustainable design and construction, waste management, sustainable procurement, carbon reduction commitment, efficient use of resources and materials, whole life costing (WLC), corporate social responsibility (CSR), and community engagement are identified by Opoku and Fortune (2015) as SC principles.

According to Ásgeirsdóttir (2013), ecological engineering, ecological design, green chemistry, GE, cradle-to-cradle, bionics, bioutilisation, biophilia, and biomimicry are those green solutions to the environmental challenges facing humanity.

Table 3.1 Principles of sustainable construction

Principles	Source
Reduce waste Improve water efficiency Reduce toxics Increase energy efficiency and renewable energy use Use environmentally preferable building materials and specifications Achieve smart growth and sustainable development Improve indoor air quality	United States Environmental Protection Agency (USEPA, 2016)
Reduce resource consumption (reduce) Reuse resources (reuse) Use recyclable resources (recycle) Protect nature (nature) Eliminate toxics (toxics) Apply life-cycle costing (economics) Focus on quality (quality)	Kibert (2013); Szydlik (2014); AlSanad (2015)
Improve sustainable site development Improve water efficiency Improve energy efficiency Conserve materials and resources Improve indoor environmental quality	United States Green Building Council (2019); Ahn et al. (2013)
Create a healthy and non-toxic environment Pursue quality in creating the built environment Minimisation of resource consumption Maximisation of resource reuse Use renewable and recyclable resources Protect the natural environment	Marhani et al. (2013); Davies and Davies (2017)
Reduce water consumption Reduce material consumption Improve indoor air quality Adopt a holistic design approach Reduce energy consumption	Office of the Federal Environmental Executive (Ahn et al., 2013)

Protection of the natural environment
Improve the quality of life
Use of renewable and recyclable resources
Profitability and competitiveness
Construct durable, functional, and quality structure
Improvement in indoor environmental quality (air,
  thermal, visual, and acoustic quality)
Respect and treat stakeholders fairly
Reduction in resource consumption (land, energy,
  materials, and water)
Ensure financial affordability
Maximisation of resource reuse
Create a healthy and non-toxic environment
Customers and clients' satisfaction and best value
Employment creation
Reduction in environmental loadings (solid waste,
  airborne emissions, liquid waste)
Adopt full-cost accounting
	Ametepey and Aigbavboa (2014)

Eliminating (material) waste Adding value to the customer	Huovila and Koskela (1998)

(Continued)

Table 3.1 (Continued)

Principles	Source
Resource efficiency	Ozgener (2006); Šaparauskas and
Energy conservation	Turskis (2006)
Pollution prevention	
System integration	
Life cycle costing	
Human quality of life (from asset to services)	Zhou and Lowe (2003)
Stakeholder partnerships	
Integration of short-term return and long-term benefits	
Maximum output with minimum input	
Value for money	
Environmental statutory compliance	Akadiri et al. (2012)
Minimise pollution	
Zero/low toxicity	
Ozone depletion potential	
Recyclable/reusable material	
Amount of likely wastage in use	
Embodied energy in material	
Environmental sound disposal options	
Impact on air quality	
Impact during harvest	
Methods of extraction of raw materials	
Maintainability	
Energy saving and thermal insulation	
Life expectancy (e.g. durability)	
Fire resistance	
Ease of construction/buildability	
Resistance to decay	
Aesthetics	
Maintenance cost	
Health and safety	
First cost	
Disposal cost	
Use of local materials	
Labour availability	
Energy conservation	Akadiri et al. (2012)
Material conservation	
Water conservation	
Land conservation	
Initial cost (purchase cost)	
Cost in use	
Recovery cost	
Protecting human health and comfort	
Protecting physical resources	
Use environmentally preferred products	The Whole Building Design
Optimise energy use	Guide (WBDG); Kubba (2012)
Optimise site potential	
Optimise operations and maintenance procedures	
Protect and conserve water	
Enhance indoor environmental quality	

Source: Author's compilation.

The Natural Step, triple bottom line (five capitals), and SC are identified by Pana-giotakopoulos and Jowitt (2004). Along similar lines, Kibert (2013) listed industrial ecology, biomimicry, eco-efficiency, the Natural Step, the precautionary princi-ple, Factor 4, and Factor 10, the biophilia hypothesis, design for the environment, ecological economics, life-cycle assessment, embodied energy, life-cycle costing, ecological rucksack, carrying capacity, construction ecology, SD concepts, and eco-efficiency as sustainability concepts in the CI.

Nawari and Ravindran (2019) listed blockchain technology (BT) and building information technology as digital tools that can be applied for a sustainable result in the CI. Others identified are nanotechnology, BT, Internet of Things (IoT), value engineering (VE), industrialised building systems (IBS), LC, building information modelling (BIM), and 3-D printing (Hussin et al., 2013; Dave et al., 2016; Lazaro et al., 2016; Heiskanen, 2017; Sakin & Kiroglu, 2017; Li et al., 2019). The emer-gence of these practices is necessitated by the need to adopt and implement eco-friendly measures in the CI while the terms describe the philosophical concepts that apply to the transition towards sustainability. A selected few of the numerous trends and practices that propel the CI towards achieving its sustainability agenda are briefly discussed coupled with their advantages.

Biomimicry

While most of the SC practices address a dimension of sustainability, others only address a subset of the dimensions. Notable among the practices is biomimicry as it seeks to achieve holistic sustainability through a balanced integration of the three dimensions of environmental, social, and economic indicators. Biomimicry advocates for a robust multi-disciplinary collaboration in the pursuit of sustainable solutions to human challenges through a comprehensive study and application of natural forms, processes, strategies, and systems. The definition by Passino (2005) presents biomimicry as a process of utilising the empirical mastery of biological systems to extract ideas from nature to proffer technological solutions. As indicated by Benyus (1997), the biomimicry concept is rooted in nature and attested to be an outstanding role model for harmony characterised by efficiency, longevity, respon-sible use of resources, and collaboration. As further affirmed by Lurie-Luke (2014), biomimicry has over the years proven to be a highly beneficial sustainability con-cept and a fertile breeding ground for disruptive innovations across all disciplines.

As a novel concept gaining increased momentum and popularity globally, bio-mimicry has the potential to sustainably transform the built environment when fully adopted and implemented. There are shreds of evidence of biomimicry engage-ment to address various global environmental concerns, aid the transition to the circular economy, improve energy management, waste treatment, pollution reduc-tion, production enhancement, waste reduction, generation of biobased energy and chemicals, and optimise sustainability holistically (Oguntona & Aigbavboa, 2017; Bockholt et al., 2019; Oguntona & Aigbavboa, 2019; Usmani et al., 2021). These benefits among numerous others can be maximally harnessed with the application of biomimicry throughout the stages of construction and a project's lifecycle.

Net Zero Energy Buildings

Globally, energy consumption is particularly significant in the built environment. Various statistics and reports from both developed and developing countries have identified building construction and usage, construction materials production, and transportation among others as energy-intensive and environmentally degrading. As reported by Pérez-Lombard et al. (2008), energy use by China alone will only take 20 years to double at an annual average growth rate of 3.7%, developed nations at 1.1%, and nations with emerging economies will increase at an annual average rate of 3.2%. Hence, the imperativeness of NZEBs as a cost-effective investment and viable panacea for minimising contamination emission levels and building energy usage (Adhikari et al., 2012). The concept as an integrated solution is aimed at reducing CO_2 emissions, fighting climate change, addressing problems of energy-saving, and reducing energy consumption (Robert & Kummert, 2012; Deng et al., 2014). To therefore define NZEBs comprehensively, an examination of the system at a minute time scale is imperative (Salom et al., 2011). From the perspective of the detailed study of Sartori et al. (2012) which dissected the terminologies of NZEBs towards a consistent definition framework, the definition by Wells et al. (2018) describes this concept as buildings with net zero GHG emissions, zero energy costs, reduced energy demands and specifically generating energy equal to usage. To this end, the various strategies and processes embraced to ensure the characteristics of NZEBs are realised are key and highly significant to optimising sustainability in the built environment. As further corroborated by Sulzakimin et al. (2020), transitioning towards SD in the built environment can be aided by the implementation of NZEBs.

Ecological Engineering

The concept and discipline of ecological engineering are similar to biomimicry as they are both underpinned by nature principles. While biomimicry promises to achieve sustainability holistically, ecological engineering solely focuses on the environmental aspect of sustainability. In China, the concept has been unceremoniously utilised for centuries but formally in the last 25 years (Mitsch et al., 1993). As a field that is rapidly developing, ecological engineering promises to inform technological advancements and solutions to present-day environmental challenges (Jørgensen, 2009). It is premised on designing eco-friendly ecosystems that incorporate human and natural environments for the benefit of both. As defined by Mitsch and Jørgensen (2003), ecological engineering entails the rejuvenation of substantially disturbed ecosystems due to human activities such as land alteration and environmental pollution and the development of new sustainable ecosystems that have both ecological and human values.

Ecological engineering offers numerous benefits to both human and natural environments with significant potential to help realise the sustainability agenda of the built environment. Ecological engineering is noted to ensure durable infrastructure, better performance structures, reduce operation and maintenance costs, and enhance energy efficiency (Barrett, 1999). The broad advantages of ecological engineering are that it enhances ecosystem health, carbon sequestration, climate regulation, production of eco-friendly and aesthetically pleasing infrastructure, pollination, air quality maintenance, pest control, sustainable water management (purification and treatment), human disease regulation, storm protection, flood control, and improve the socio-economic status of the people (Costanza, 2012; Poff et al., 2016). In summary, it can be inferred that ecological engineering offers a trio of environmental, economic, and social benefits to the built environment.

Nanotechnology

Nanotechnology is one of the technologies gaining global popularity in the CI with significant potential for enhancing sustainability. This new concept has gained visibility and increased patronage across several fields. However, there is no definite definition of the concept of nanotechnology from the review of relevant literature across different fields of study. Definitions found are based on the diverse base of scientists responding to novel research observations in the nanotechnology concept to fit their different research niche areas (Balogh, 2010). As defined by McNeil (2005) and Hulla et al. (2015), nanotechnology is the mastery and sway of matter at measurements between 1 and 100 nm where distinctive phenomena facilitate new applications. It is also described by Tiwari et al. (2013) as the 'engineering of functional systems at the molecular state'. Nanotechnology is perceived as the next industrial revolution and the future of advanced development as it promotes new products, devices, and materials that exhibit divergent attributes (Gilman, 2001; Kumar & Jee, 2013). It has the potential to proffer solutions to various environmental and technological issues in the domains of water treatment, solar energy conversion, medicine, catalysis, drug supply chain, modern agriculture, engineering, CI, food processing, and packaging (Sailaja et al., 2014; Sekhon, 2014; Rao et al., 2015; Lazaro et al., 2016; Singh et al., 2017; Nasrollahzadeh et al., 2019).

A major area of the CI in which nanotechnology is perceived to be highly significant is energy and material resources. The concept of nanotechnology is important in this sector due to the unique role that construction materials and products play in determining the severity of the impact of buildings and infrastructure on the environment. The available evidence in the study of Abdin et al. (2018) suggested that the application of nanotechnology plays an important role in achieving energy efficiency. Along similar lines, Rao et al. (2015) indicated that nanotechnology is an influential concept in the field of civil engineering, especially in making concrete as it helps in producing more durable, strong, and eco-friendly construction materials. Overall, nanotechnology can enhance SC practices by optimising the functionality of traditional construction materials, create novel construction material economy, reduce material carbon emissions, reduce energy consumption, addresses urban sustainability challenges, minimise material environmental impacts, introduce economically viable materials, increase infrastructure durability, create sustainable infrastructure, reduce raw materials consumption, support materials with high recyclable content, and propels the need for more SD (Oke et al., 2017).

Additive Manufacturing/3-D Printing

Additive manufacturing (AM) is perceived to possess numerous advantages over conventional subtractive or formative manufacturing. It is regarded as one of the advanced manufacturing concepts with unprecedented development potential which has rapidly evolved in the past 30 years (Lu et al., 2015). According to Sames et al. (2016) and Mahamood et al. (2019), it is a manufacturing procedure for fabrication using 3-D computer-aided design models by adding materials in a process that proceeds layer by layer. Also referred to as 3-D printing, AM proffers a time-efficient and cost-effective means of producing low-volume customised materials and products with advanced functionality, attributes, and complex geometries (Huang et al., 2015). Despite its complex processes, AM unlocks design potential with numerous benefits and sustainable footprints (Petrovic et al., 2011; Bikas et al., 2016; Machado et al., 2019).

The identification of AM as one of the most beneficial and propitious technological concepts is that it optimises sustainability in the built environment (Hao et al., 2010; Peng et al., 2018). It can be applied to component manufacturing and rapid prototyping across various sectors (Attaran, 2017). With an exponentially increased application for building construction over the years, the potential benefits of AM from a sustainability perspective cut across material and product lifecycles. It facilitates and enables the production (in little to massive quantities) of custom-made materials, enables the construction of complex geometries, increases job site safety, enhances cost-effectiveness, produces materials with astonishing properties, reduces material waste on construction sites, increases the overall quality of building projects, improves job site productivity, and project sustainability among others (Mitchell et al., 2018; Akhnoukh, 2021; Singh & Agrawal, 2021).

Internet of Things (IoT)

The Internet of Things (IoT) concept is one of the core technologies of the 4IR. It has attracted widespread and increasing global attention due to its prospects and application across various disciplines. The present 4IR era has seen the widespread application of IoT in building construction, defence and security, healthcare, agriculture, supply chain, and logistics management to mention a few. As defined by Sethi and Sarangi (2017), it is a paradigm where objects, integrated with sensors, actuators, and processors, establish interconnected communication to accomplish specific and significant objectives. This concept encompasses smart systems and devices intelligently connected to communicate and interact with themselves and other objects, infrastructure, and the environment (Dudhe et al., 2017). IoT has significantly transformed and elevated people's lifestyles to new and unimaginable heights through its ambiently intelligent, context-aware, and predominant devices (Whitmore et al., 2015). As indicated by Junaidi (2015), these devices include wireless sensor networks, radio frequency identification (RFID), media sensors, and other smart objects which are optimised to ensure seamless interaction between the devices connected to the internet network and humans.

One of the greatest importance of IoT lies in its ability to generate, analyse, and make informed decisions on information about connected devices (Ismail, 2019), a feature that has the potential to optimise sustainability in the built environment. As highlighted by several researchers, IoT can address challenges in energy security, ensure informed, sustainable, and robust resource management in urban and rural communities, and trigger the creation of value networks and new markets, (Nasiri et al., 2019, Salam, 2020a; Salam, 2020b). Other benefits include early detection of flaws in construction phases, improved project handling, prompt maintenance of equipment and machinery, minimised project delay, provision of up-to-date information for better decision-making, prevention of accidents on construction job sites, enhanced health and safety on construction sites, support cost and time efficiency, and improved construction management performance (Arslan et al., 2019; Maqbool et al., 2023). These benefits of IoT among others have the potential to help the CI achieve its sustainability agenda.

Building Information Modeling (BIM)

Regarded as a process and technology, BIM is a novel paradigm in the built environment that enables the integration of the roles of all construction stakeholders (Azhar et al., 2012). Along similar lines, Succar et al. (2007) indicated that BIM is composed of three interconnected knowledge nodes of policy, process, and technology. These nodes identify, capture, and represent BIM interactions for optimal and sustainable results. BIM consists of a cohesive framework of technologies, processes, and policies for establishing a methodology to effectively manage crucial design and data of building projects in a digital format throughout the lifecycle of the project.

When adopted and implemented, BIM has great potential in the design and development of green buildings. It allows for an appropriate, real-time, and accurate exchange of information, increases total project quality, improves scheduling, minimises total project costs and contingencies, provides accurate quantity take-offs, streamlines and eases the green building certification process, and optimises building designs to improve their sustainability (Azhar et al., 2011; Bynum et al., 2013; Demian & Walters, 2014; Liu et al., 2015). Other benefits of BIM for optimising sustainability in the built environment include enhancement of project closeout, risk reduction of project data, assistance in controlling subcontractors, improved project collaboration, assistance in risk identification and management, improvement in visualisation of construction details, improvement of safety features, reduction of construction errors, reworks and wastes, increased workforce effectiveness, clash detection, dispute resolution, and optimising clients satisfaction (Ibrahim et al., 2019).

Value Engineering

The concept of VE is gaining unprecedented popularity across all sectors globally (Oke & Aghimien, 2018). Terminologies such as value assurance, value improvement, value methodology, value control, value analysis, value planning, and value management have been found across different literature to be used interchangeably with VE (Chavan, 2013; Oke & Aigbavboa, 2017). Due to the high risks associated with large infrastructural projects, VE is imperative as it offers a methodically innovative approach to identifying unnecessary costs in projects (Tohidi, 2011; Mansour & Abueusef, 2015). It is defined as the 'creative and organised effort, which analyses the requirements of a project to achieve the essential functions at the lowest total costs over the life of the project' (Rane & Attarde, 2016). VE has now become an integral part of construction management as a tool to improve processes and services vis a vis project quality, cost, and time. This has made the concept another cogent enabler of sustainability in the built environment.

There are several benefits accrued to the adoption and implementation of VE, especially in the construction sector. According to Danku and Antwi (2020), the benefits of VE are to ensure contract compliance, effective construction delivery, and project sustainability. Other benefits that VE offers include: encouraging the use of local construction materials, enhancing the quality performance of construction projects, reducing project abandonment, promoting project adaptability and flexibility, aiding decision-making, promoting adaptability and flexibility, enhancing economic investment, aiding high technical advancement, ensuring technological advancement, enhancing competitive edge for the contractor, eliminating unnecessary design, aiding healthy teamwork relationship among construction professionals, encouraging efficient conflict management, and promoting effective project delivery services (Oke & Ogunsemi, 2011). These benefits among others will ensure the optimal function of VE to aid sustainability in the built environment.

Green Engineering

The materials and products specified and employed on construction projects and infrastructure delivery dictate and determine the eco-friendliness of such projects. The continued degrading and negative environmental impacts of buildings are traceable to the high number of conventional and unsustainable materials and products used. This, therefore, reiterates the significance of green materials and products (enabled by the field of GE) in the sustainability agenda of the built environment. According to Navinchandra (1991), GE is described as the process of examining product and process design with a major emphasis on its eco-friendly attributes without compromising commercial feasibility and quality. GE is precipitated on the key elements of beyond-the-plant boundary considerations, environmentally conscious design, and environmental literacy (Shonnard et al., 2003). Using lifecycle thinking, conserving natural resources, ensuring safe material and energy inputs, engineering processes and products holistically, using system analysis, integrating environmental impact assessment tools, striving to prevent waste, minimising natural resources depletion, developing and applying engineering solutions, creating engineering solutions beyond current technologies, and actively engaging stakeholders and communities in the development of engineering solutions are presented by Abraham and Nguyen (2003) as GE principles. Based on its inherent principles, embracing and incorporating GE at the pre-design and design stages of materials and products has the potential to lead to SD (Hesketh et al., 2004). It is noteworthy that GE utilises renewable natural resources, and safe solvents to design and make materials and products that alter the waste pattern, minimise pollution, and does not deteriorate the earth. Owing to the important role of green materials and products in achieving sustainable buildings and infrastructure, GE can be agreed to be inevitable in the design and manufacture of these products.

Lean Construction

The complexities, high resource consumption, emission and waste generation attributes of the CI have attracted global attention. This led to the need to embrace lean thinking, which is a new way of construction project management as its techniques possess the capacity to influence the bottom line of projects (Howell & Ballard, 1998; Salem et al., 2006). LC is therefore the application of lean thinking in the construction sector. There are numerous studies with diverse definitions of LC making the task of defining the concept an uphill one. However, from the study of Mossman (2018) which dissected several definitions, LC can be stated as the 'practical collection of theories, principles, axioms, techniques, and ways of thinking that together and severally can help individuals and teams improve the processes and systems within' construction projects.

What makes LC significant to the sustainability agenda of the built environment is that it is an excellent tool for eliminating waste, managing construction processes, and achieving the project's goals (Marhani et al., 2013). LC also improves the safety performance of projects, prevents pollution, improves the whole-life cost of construction projects, minimises resource depletion, enhances the efficient administration of materials, reduces waste, promotes cost and time efficiency, maximises value, and improves sustainability (Marhani et al., 2012; Bajjou et al., 2017; Radhika & Sukumar, 2017; Akinradewo et al., 2018). Ultimately, the concept of LC has over the years proven to be a potent SC practice when embraced and fully implemented.

Blockchain Technology

BT is one of the technologies that characterises the 4IR era. While it is a concept with a significant imprint in the information and communication technology (ICT) sector, there are now records of BT adoption and implementation in other disciplines and sectors. When attempting to comprehend the concept of blockchain in the context of electronic money, it can be understood as a combination of two fundamental elements: 'blocks' and 'chains'. The term 'block' refers to the protocols governing transaction data, while 'chain' pertains to the protocols dictating the interconnectedness of these blocks (Wang & Su, 2020). From this word segmentation, BT is regarded as a distributed ledger technology that enables the recording of various operations and transactions in a sequential chain of blocks, all without the need for intermediaries (Vaigandla et al., 2023). It is a decentralised, open, and immutable distributed public ledger that facilitates transactions without the need for a trusted third party (Karthika & Jaganathan, 2019). One of the foremost implementations of BT is a cryptocurrency named Bitcoin where its protocol allows individuals to effortlessly transfer money between each other, irrespective of their banking affiliations (Swan, 2017). BT is poised to revolutionise various sectors of the economy as a disruptive force as it allows for an open and decentralised society.

There are several benefits of BT when adopted and implemented in the built environment with great potential to aid the achievement of SD. The benefits inferred from the study of Abu-Elezz et al. (2020) show that BT can help with monitoring, information exchange, management, tracking, security, and authorisation purposes. However, the study by Perera et al. (2020) noted that BT has numerous benefits to offer when applied in property management, file sharing for document management, construction supply chain management, asset management, embodied carbon management, water trading, construction management, energy management, building maintenance system, payment management, incorporation with BIM as a procurement solution, and waste management areas of the built environment. Also, BT is believed to help reduce transaction costs, minimise carbon footprints, enhance communication and performance along the construction supply chain, contribute to a circular economy, enhance the visibility of sustainability practices, and improve environmental practices (Rejeb & Rejeb, 2020; Najjar, 2021, Upadhyay et al., 2021). These benefits and many others are pointers to the capabilities of BT to drive the sustainability agenda of the built environment.

Industrialised Building System (IBS)

The IBS has gained popularity and global acceptability owing to the numerous benefits it offers. While some countries have fully embraced and implemented this concept in housing and infrastructure delivery, others only implement it for some elements of the building block. Throughout the global body of knowledge, several definitions are found to describe IBS. Considering the various definitions examined by Azman et al. (2018), IBS can be described as a construction method where the building elements are fabricated offsite/onsite using detailed standard measurements and shapes, and thereafter transported to the site for assembling into a structure. Also, other terminologies are found to have the same contextual underpinning as the IBS concept. For example, terminologies such as industrialised construction and fabrication, prefabrication construction, prefabrication, modular construction, offsite manufacturing, modern method of construction, pre-assembly, modular integrated construction, permanent modular construction, offsite construction, prefabricated prefinished volumetric construction, system building, and panelised building system are used interchangeably with IBS (Kamar et al., 2011; Thai et al., 2020). All these terms coupled with IBS represent a sustainable and more industrialised method of construction compared to the popular conventional method.

IBS is affirmed to possess the propensity to fulfil the goals of SC as a result of the principles of sustainability embedded in the concept which helps in maintaining the balance between construction and the environment (Zabihi et al., 2013; Musa et al., 2014). Benefits offered by the implementation of IBS for sustainability include less site disturbance from equipment, suppliers, and workers, inventory control, ability to service remote locations, reduced demand for raw materials, minimised energy demand, flexible, movable, and reusable building elements, improved health and safety on site, reduced site accidents, reduced project delays, eased refurbishment and renovation work, improved air quality, minimised materials wastage, and faster and efficient construction projects (Musa et al., 2016). Other benefits are less labour dispute on site, reduced labour cost, fewer defects, optimised use of construction materials, cleaner and neater construction site, reduced project duration, improved quality of projects, and materials efficiency (Bari et al., 2012; Wong & Lau, 2014; Ayorinde et al., 2021). These benefits are what the IBS concept offers to aid the transition of the built environment to a truly sustainable state.

Summary

This chapter presented a brief state of the CI and introduced the concept of sustainability within the sector. The various principles that underpin the concept of SC are also presented. This chapter also presents the SC concepts, trends, and practices that are proliferated and administered to aid the transition of the built environment to a sustainable state. Eleven of these trends and practices namely Additive Manufacturing/3D printing, Ecological Engineering, VE, Biomimicry, Green Engineering, LC, Net Zero Energy Building, IBS, BIM, Nanotechnology, Internet of Things, and BT are highlighted and discussed. This chapter also presented the benefits of these SC trends and practices to show their contribution to the sustainability agenda of the built environment. The next chapter presents an overview of the green building paradigm by addressing the goals, benefits, barriers, drivers, and roles of green building collaborative networks towards achieving SD in the construction sector.

References

Abdin, A. R., El Bakery, A. R., & Mohamed, M. A. (2018). The role of nanotechnology in improving the efficiency of energy use with a special reference to glass treated with nanotechnology in office buildings. *Ain Shams Engineering Journal, 9*(4), 2671–2682.

Abdul Nifa, F. A., Mohd Nawi, M. N., Osman, W. N., & Abdul Rahim, S. (2015). Towards development of sustainable design in Malaysian university campus: A preliminary framework for universiti utara Malaysia. *Jurnal Teknologi, 77*(5), 43–49.

Abraham, M. A., & Nguyen, N. (2003). "Green engineering: Defining the principles" – Results from the Sandestin conference. *Environmental Progress, 22*(4), 233–236.

Abu-Elezz, I., Hassan, A., Nazeemudeen, A., Househ, M., & Abd-Alrazaq, A. (2020). The benefits and threats of blockchain technology in healthcare: A scoping review. *International Journal of Medical Informatics, 142*, 104246.

Adhikari, R. S., Aste, N., Del Pero, C., & Manfren, M. (2012). Net zero energy buildings: Expense or investment? *Energy Procedia, 14*, 1331–1336.

Ahn, Y. H., Pearce, A. R., Wang, Y., & Wang, G. (2013). Drivers and barriers of sustainable design and construction: The perception of green building experience. International *Journal of Sustainable Building Technology and Urban Development, 4*(1), 35–45.

Akadiri, P. O., Chinyio, E. A., & Olomolaiye, P. O. (2012). Design of a sustainable building: A conceptual framework for implementing sustainability in the building sector. *Buildings, 2*(2), 126–152.

Akhnoukh, A. K. (2021). Advantages of contour crafting in construction applications. *Recent Patents on Engineering, 15*(3), 294–300.

Akinradewo, O., Oke, A., Aigbavboa, C., & Ndalamba, M. (2018). Benefits of adopting lean construction technique in the South African construction industry. In *Proceedings of the International Conference on Industrial Engineering and Operations Management, 29 October to 1 November 2018, Pretoria, South Africa.*

AlSanad, S. (2015). Awareness, drivers, actions, and barriers of sustainable construction in Kuwait. *Procedia Engineering, 118*, 969–983.

Alves, T. C. L., Milberg, C., & Walsh, K. D. (2012). Exploring lean construction practice, research, and education. *Engineering, Construction and Architectural Management, 19*(5), 512–525.

Ametepey, S. O., & Aigbavboa, C. (2014). Practitioners perspectives for the implementation of sustainable construction in the Ghanaian construction industry. *Proceedings of the DII-2014 Conference on Infrastructure Investments in Africa*, 114–124.

Arslan, V., Ulubeyli, S., & Kazaz, A. (2019). The use of Internet of Things in the construction industry. *UEMK 2019 Proceedings Book 24/25 October 2019 Gaziantep University, Turkey*, 501–510.

Ásgeirsdóttir, S. A. (2013). *Biomimicry in Iceland: Present status and future significance.* University of Iceland.

Ashour, A., Bragança, L., & Almeida, M. G. D. (2015). Promoting sustainability as a strategy to mitigate the effects of economic downturn on the construction industry. *Euro Elecs, 3,* (Chapter 12: Policies and strategies for a sustainable built environment) 1745–1752.

Attaran, M. (2017). The rise of 3-D printing: The advantages of additive manufacturing over traditional manufacturing. *Business Horizons, 60*(5), 677–688.

Ayorinde, E. O., Ngcobo, N., & Kasenge, M. (2021). Benefits of Adopting Industrialized Building Systems in the South African Construction Industry. In *2021 International Conference on Advanced Enterprise Information System (AEIS)* (pp. 44–47). IEEE.

Azhar, S., Carlton, W. A., Olsen, D., & Ahmad, I. (2011). Building information modeling for sustainable design and leed® rating analysis. *Automation in Construction, 20*(2), 217–224.

Azhar, S., Khalfan, M., & Maqsood, T. (2012). Building information modeling (BIM): Now and beyond. *Australasian Journal of Construction Economics and Building, 12*(4), 15–28.

Azman, N. S., Ramli, M. Z., & Zawawi, M. H. (2018). Factors affecting quality management of construction project using industrialized building system: A review. *International Journal of Engineering and Technology, 7*(4), 307–311.

Bajjou, M. S., Chafi, A., & En-Nadi, A. (2017). The potential effectiveness of lean construction tools in promoting safety on construction sites. *International Journal of Engineering Research in Africa, 33*, 179–193.

Balogh, L. P. (2010). Why do we have so many definitions for nanoscience and nanotechnology? *Nanomedicine: Nanotechnology, Biology, and Medicine, 6*(3), 397–398.

Bari, N. A. A., Abdullah, N. A., Yusuff, R., Ismail, N., & Jaapar, A. (2012). Environmental awareness and benefits of industrialized building systems (IBS). *Procedia-Social and Behavioral Sciences, 50*, 392–404.

Barrett, K. R. (1999). Ecological engineering in water resources: The benefits of collaborating with nature. *Water International, 24*(3), 182–188.

Benyus, J. M. (1997). *Biomimicry: Innovation inspired by nature* (Adobe Digital ed.). Australia: HarperCollins.

Bikas, H., Stavropoulos, P., & Chryssolouris, G. (2016). Additive manufacturing methods and modelling approaches: A critical review. *The International Journal of Advanced Manufacturing Technology, 83*, 389–405.

Blizzard, J. L., & Klotz, L. E. (2012). A framework for sustainable whole systems design. *Design Studies, 33*(5), 456–479.

Bockholt, M. T., Kristensen, J. H., Wæhrens, B. V., & Evans, S. (2019). Learning from the nature: Enabling the transition towards circular economy through biomimicry. In *2019 IEEE International Conference on Industrial Engineering and Engineering Management (IEEM)* (pp. 870–875). IEEE.

Bynum, P., Issa, R. R., & Olbina, S. (2013). Building information modeling in support of sustainable design and construction. *Journal of Construction Engineering and Management, 139*(1), 24–34.

Chavan, A. J. (2013). Value engineering in construction industry. *International Journal of Application or Innovation in Engineering & Management, 2*(12), 18–26.

Costanza, R. (2012). Ecosystem health and ecological engineering. *Ecological Engineering, 45*, 24–29.

Danku, J. C., & Antwi, P. A. (2020). Perceived benefits of using value engineering on road projects in Ghana. *World Journal of Engineering and Technology, 8*(2), 217–236.

Dave, B., Kubler, S., Främling, K., & Koskela, L. (2016). Opportunities for enhanced lean construction management using Internet of Things standards. *Automation in Construction, 61*, 86–97.

Davies, O., & Davies, I. (2017). Barriers to implementation of sustainable construction techniques. *MAYFEB Journal of Environmental Science, 2*, 1–9.

Demian, P., & Walters, D. (2014). The advantages of information management through building information modelling. *Construction Management and Economics, 32*(12), 1153–1165.

Deng, S., Wang, R. A., & Dai, Y. J. (2014). How to evaluate performance of net zero energy building–A literature research. *Energy, 71*, 1–16.

Dudhe, P. V., Kadam, N. V., Hushangabade, R. M., & Deshmukh, M. S. (2017). Internet of Things (IoT): An overview and its applications. In *2017 International Conference on Energy, Communication, Data Analytics and Soft Computing (ICECDS)* (pp. 2650–2653). IEEE.

Gan, X., Zuo, J., Ye, K., Skitmore, M., & Xiong, B. (2015). Why sustainable construction? Why not? An owner's perspective. *Habitat International, 47*, 61–68.

Gibberd, J. (2002). The sustainable building assessment tool assessing how buildings can support sustainability in developing countries. In *Built Environment Professions Convention, 1–3 May 2002, Johannesburg, South Africa*.

Gilman, J. (2001) Nanotechnology, *Materials Research Innovations*, 5:1, 12–14. https://doi.org/10.1007/s100190100124

Goh, C. S. (2014). *Development of a capability maturity model for sustainable construction*. The University of Hong Kong.

Hao, L., Raymond, D., Strano, G., & Dadbakhsh, S. (2010). Enhancing the sustainability of additive manufacturing. In *5th International Conference on Responsive Manufacturing-Green Manufacturing (ICRM 2010)* (pp. 390–395). IET.

Heiskanen, A. (2017). The technology of trust: How the Internet of Things and blockchain could usher in a new era of construction productivity. *Construction Research and Innovation, 8*(2), 66–70.

Hesketh, R. P., Slater, C. S., Savelski, M. J., Hollar, K., & Farrell, S. (2004). A program to help in designing courses to integrate green engineering subjects. *International Journal of Engineering Education, 20*(1), 113–123.

Howell, G., & Ballard, G. (1998). Implementing lean construction: understanding and action. In *Proceedings of 6th Annual Conference International Group for Lean Construction.*

Huang, Y., Leu, M. C., Mazumder, J., & Donmez, A. (2015). Additive manufacturing: Current state, future potential, gaps and needs, and recommendations. *Journal of Manufacturing Science and Engineering, 137*(1).

Hulla, J. E., Sahu, S. C., & Hayes, A. W. (2015). Nanotechnology: History and future. *Human & Experimental Toxicology, 34*(12), 1318–1321.

Huovila, P., & Koskela, L. (1998). Contribution of the principles of lean construction to meet the challenges of sustainable development. *6th Annual Conference of the International Group for Lean Construction,* Guaruja, São Paulo, Brazil (pp. 1–11). IGLG.

Hussin, J. M., Rahman, I. A., & Memon, A. H. (2013). The way forward in sustainable construction: Issues and challenges. *International Journal of Advances in Applied Sciences, 2*(1), 15–24.

Ibrahim, H. S., Hashim, N., & Jamal, K. A. A. (2019, November). The potential benefits of building information modelling (BIM) in construction industry. In *IOP Conference Series: Earth and Environmental Science* (Vol. 385, No. 1, p. 012047). IOP Publishing.

Ismail, Y. (2019). Introductory chapter: Internet of Things (IoT) importance and its applications. In *Internet of Things (IoT) for automated and smart applications.* IntechOpen.

Jørgensen, S. E. (Ed.). (2009). *Applications in ecological engineering.* Academic Press.

Junaidi, A. (2015). Internet of things, sejarah, teknologi dan penerapannya. *Jurnal Ilmiah Teknologi Infomasi Terapan, 1*(3), 62–66.

Kamar, A. M., Abd Hamid, Z., & Azman, N. A. (2011). Industrialized building system (IBS): Revisiting issues of definition and classification. *International Journal of Emerging Sciences, 1*(2), 120.

Karthika, V., & Jaganathan, S. (2019). A quick synopsis of blockchain technology. *International Journal of Blockchains and Cryptocurrencies, 1*(1), 54–66.

Kibert, C. J. (2013). *Sustainable construction: Green building design and delivery.* John Wiley & Sons.

Kibert, C. J., & Fenner, A. E. (2017). *Sustainable manufacturing: Design and construction strategies for manufactured construction.* Florida, US: University of Florida.

Kibert, C. J., Sendzimir, J., & Guy, B. (2000). Construction ecology and metabolism: Natural system analogues for a sustainable built environment. *Construction Management & Economics, 18*(8), 903–916.

Kubba, S. (2012). *Handbook of green building design and construction: LEED, BREEAM, and green globes.* USA: Butterworth-Heinemann.

Kumar, A., & Jee, M. (2013). Nanotechnology: A review of applications and issues. *International Journal of Innovative Technology and Exploring Engineering (IJITEE), 3*(4), 1–2.

Lazaro, A., Yu, Q. L., & Brouwers, H. (2016). Nanotechnologies for sustainable construction. *Sustainability of construction materials,* 55–78, Elsevier.

Lazaro, A., Yu, Q. L., & Brouwers, H. J. H. (2016). Nanotechnologies for sustainable construction. In *Sustainability of construction materials* (pp. 55–78). Woodhead Publishing.

Li, J., Greenwood, D., & Kassem, M. (2019). Blockchain in the built environment and construction industry: A systematic review, conceptual models and practical use cases. *Automation in Construction, 102*, 288–307.

Li, W., & Li, H. (2019). Research of green architecture – Take Chinese traditional cave dwellings as an example. *IOP Conference Series: Earth and Environmental Science, 310*(2), 022054.

Liu, S., Meng, X., & Tam, C. (2015). Building information modeling based building design optimization for sustainability. *Energy and Buildings, 105*, 139–153.

Lu, B., Li, D., & Tian, X. (2015). Development trends in additive manufacturing and 3D printing. *Engineering, 1*(1), 85–89.

Lurie-Luke, E. (2014). Product and technology innovation: What can biomimicry inspire? *Biotechnology Advances, 32*(8), 1494–1505.

Machado, C. G., Despeisse, M., Winroth, M., & da Silva, E. H. D. R. (2019). Additive manufacturing from the sustainability perspective: Proposal for a self-assessment tool. *Procedia CIRP, 81*, 482–487.

Mahamood, R. M., Akinlabi, S. A., Shatalov, M., Murashkin, E. V., & Akinlabi, E. T. (2019). Additive manufacturing/3D printing technology: A review. *Annals of "Dunarea De Jos" University of Galati. Fascicle XII, Welding Equipment and Technology, 30*, 51–58.

Mansour, A. K., & Abueusef, M. (2015). Value engineering in developing countries. In *Proceedings of the International Conference Data Mining, Civil and Mechanical Engineering (ICDMCME'2015)* (pp. 1–2).

Maqbool, R., Saiba, M. R., & Ashfaq, S. (2023). Emerging Industry 4.0 and Internet of Things (IoT) technologies in the Ghanaian construction industry: Sustainability, implementation challenges, and benefits. *Environmental Science and Pollution Research, 30*(13), 37076–37091.

Marhani, M. A., Adnan, H., & Ismail, F. (2013). OHSAS 18001: A pilot study of towards sustainable construction in Malaysia. *Procedia – Social and Behavioral Sciences, 85*, 51–60.

Marhani, M. A., Jaapar, A., & Bari, N. A. A. (2012). Lean construction: Towards enhancing sustainable construction in Malaysia. *Procedia-Social and Behavioral Sciences, 68*, 87–98.

Marhani, M. A., Jaapar, A., Bari, N. A. A., & Zawawi, M. (2013). Sustainability through lean construction approach: A literature review. *Procedia-Social and Behavioral Sciences, 101*, 90–99.

McNeil, S. E. (2005). Nanotechnology for the biologist. *Journal of Leukocyte Biology, 78*(3), 585–594.

Mitchell, A., Lafont, U., Hołyńska, M., & Semprimoschnig, C. J. A. M. (2018). Additive manufacturing – A review of 4D printing and future applications. *Additive Manufacturing, 24*, 606–626.

Mitsch, W. J., & Jørgensen, S. E. (2003). Ecological engineering: A field whose time has come. *Ecological Engineering, 20*(5), 363–377.

Mitsch, W. J., Yan, J., & Cronk, J. K. (1993). Ecological engineering – Contrasting experiences in China with the West. *Ecological Engineering, 2*(3), 177–191.

Mossman, A. (2018). What is lean construction: another look-2018. In *26th Annual Conference of the International Group for Lean Construction* (pp. 1240–1250).

Mousa, A. (2015). A business approach for transformation to sustainable construction: An implementation on a developing country. *Resources, Conservation and Recycling, 101*, 9–19.

Musa, M. F., Mohammad, M. F., Mahbub, R., & Yusof, M. R. (2014). Enhancing the quality of life by adopting sustainable modular industrialised building system (IBS) in the Malaysian construction industry. *Procedia-Social and Behavioral Sciences, 153*, 79–89.

Musa, M. F., Yusof, M. R., Mohammad, M. F., & Samsudin, N. S. (2016). Towards the adoption of modular construction and prefabrication in the construction environment: A case study in Malaysia. *Journal of Engineering and Applied Sciences, 11*(13), 8122–8131.

Najjar, M. (2021). Blockchain technology for reinforcing sustainability practices across complex multi-tier supply networks. In A.M. Musleh Al-Sartawi, A. Razzaque, & M. M. Kamal (Eds.), *Artificial intelligence systems and the internet of things in the digital era.* EAMMIS 2021. Lecture Notes in Networks and Systems, Vol. 239. Springer, Cham.

Nasiri, H., Nasehi, S., & Goudarzi, M. (2019). Evaluation of distributed stream processing frameworks for IoT applications in Smart Cities. *Journal of Big Data, 6*, 1–24.

Nasrollahzadeh, M., Sajadi, S. M., Sajjadi, M., & Issaabadi, Z. (2019). An introduction to nanotechnology. In *Interface science and technology,* 28, 1–27, Elsevier.

Navinchandra, D. (1991). Design for environmentability. In *International Design Engineering Technical Conferences and Computers and Information in Engineering Conference* (Vol. 7477, pp. 119–125). American Society of Mechanical Engineers.

Nawari, N. O., & Ravindran, S. (2019). Blockchain and building information modelling (BIM): Review and applications in post-disaster recovery. *Buildings, 9*(149), 1–32.

Oguntona, O. A., & Aigbavboa, C. O. (2017). Biomimicry principles as evaluation criteria of sustainability in the construction industry. *Energy Procedia, 142*, 2491–2497.

Oguntona, O. A., & Aigbavboa, C. O. (2019). Biomimicry interventions for addressing global environmental challenges. In *Creative Construction Conference 2019* (pp. 95–101). Budapest University of Technology and Economics.

Oke, A. E., & Aghimien, D. O. (2018). Drivers of value management in the Nigerian construction industry. *Journal of Engineering, Design and Technology, 16*(2), 270–284.

Oke, A. E., & Aigbavboa, C. O. (2017). *Sustainable value management for construction projects* (pp. 75–86). Switzerland: Springer.

Oke, A. E., Aigbavboa, C. O., & Semenya, K. (2017). Energy savings and sustainable construction: Examining the advantages of nanotechnology. *Energy Procedia, 142*, 3839–3843.

Oke, A. E., & Ogunsemi, D. R. (2011). Value management in the Nigerian construction industry: Militating factors and the perceived benefits. In *Second International Conference on Advances in Engineering and Technology* (pp. 353–359).

Opoku, A., & Fortune, C. (2015). Current practices towards achieving sustainable construction project delivery in the UK. *The International Journal of Environmental, Cultural, Economic, and Social Sustainability: Annual Review, 10*(1), 41–57.

Ozgener, O. (2006). A small wind turbine system (SWTS) application and its performance analysis. *Energy Conversion and Management, 47*(11), 1326–1337.

Panagiotakopoulos, P. D., & Jowitt, P. W. (2004). Representing and assessing sustainability in construction. In F. Khosrowshahi (Ed.), *20th Annual ARCOM Conference, 1–3 September 2004, Heriot Watt University* (Vol. 2, pp. 1305–1311). Association of Researchers in Construction Management.

Passino, K. M. (2005). *Biomimicry for optimization, control, and automation.* Springer Science & Business Media.

Peng, T., Kellens, K., Tang, R., Chen, C., & Chen, G. (2018). Sustainability of additive manufacturing: An overview on its energy demand and environmental impact. *Additive Manufacturing, 21*, 694–704.

Perera, S., Nanayakkara, S., Rodrigo, M. N. N., Senaratne, S., & Weinand, R. (2020). Blockchain technology: Is it hype or real in the construction industry? *Journal of Industrial Information Integration, 17*, 100125.

Pérez-Lombard, L., Ortiz, J., & Pout, C. (2008). A review on buildings energy consumption information. *Energy and Buildings, 40*(3), 394–398.

Petrovic, V., Vicente Haro Gonzalez, J., Jordá Ferrando, O., Delgado Gordillo, J., Ramón Blasco Puchades, J., & Portolés Griñan, L. (2011). Additive layered manufacturing: Sectors of industrial application shown through case studies. *International Journal of Production Research, 49*(4), 1061–1079.

Plank, R. (2008). The principles of sustainable construction. *The IES Journal Part A: Civil & Structural Engineering, 1*(4), 301–307.

Poff, N. L., Brown, C. M., Grantham, T. E., Matthews, J. H., Palmer, M. A., Spence, C. M., & Baeza, A. (2016). Sustainable water management under future uncertainty with eco-engineering decision scaling. *Nature Climate Change, 6*(1), 25–34.

Presley, A., & Meade, L. (2010). Benchmarking for sustainability: An application to the sustainable construction industry. *Benchmarking: An International Journal, 17*(3), 435–451.

Radhika, R., & Sukumar, S. (2017). An overview of the concept of lean construction and the barriers in its implementation. *International Journal of Engineering Technologies and Management Research, 4*(3), 13–26.

Rane, N. L., & Attarde, P. M. (2016). Application of value engineering in construction projects. *International Journal of Engineering and Management Research (IJEMR), 6*(1), 25–29.

Rao, N. V., Rajasekhar, M., Vijayalakshmi, K., & Vamshykrishna, M. (2015). The future of civil engineering with the influence and impact of nanotechnology on properties of materials. *Procedia Materials Science, 10*, 111–115.

Rejeb, A., & Rejeb, K. (2020). Blockchain and supply chain sustainability. *Logforum, 16*(3).

Robert, A., & Kummert, M. (2012). Designing net-zero energy buildings for the future climate, not for the past. *Building and Environment, 55*, 150–158.

Said, I., Osman, O., Shafiei, M. W., Rashideh, W. M. A., & Kooi, T. K. (2009). *Modeling of construction firm's sustainability*. ICCI, 1–12.

Sailaja, A. K., Reddy, A. S., Sreelola, V., Swathi, P., & Vineela, C. (2014). Nanotechnology-an overview. *Journal of Pharmacy and Nutrition Sciences, 4*, 246–254.

Sakin, M., & Kiroglu, Y. C. (2017). 3D printing of buildings: Construction of the sustainable houses of the future by BIM. *Energy Procedia, 134*, 702–711.

Salam, A. (2020a). Internet of things for sustainable community development: Introduction and overview. *Internet of things for sustainable community development: Wireless communications, sensing, and systems*, 1–31. Springer, Cham.

Salam, A. (2020b). Internet of things in sustainable energy systems. *Internet of things for sustainable community development: Wireless communications, sensing, and systems*, 183–216. Springer, Cham.

Saleh, M. S., & Alalouch, C. (2015). Towards sustainable construction in Oman: Challenges & opportunities. *Procedia Engineering, 118*, 177–184.

Salem, O., Solomon, J., Genaidy, A., & Minkarah, I. (2006). Lean construction: From theory to implementation. *Journal of Management in Engineering, 22*(4), 168–175.

Salom, J., Widén, J., Candanedo, J., Sartori, I., Voss, K., & Marszal, A. (2011). Understanding net zero energy buildings: evaluation of load matching and grid interaction indicators. In *Proceedings of Building Simulation* (Vol. 6, pp. 2514–2521).

Sameh, S. H. (2014). Promoting earth architecture as a sustainable construction technique in Egypt. *Journal of Cleaner Production, 65*, 362–373.

Sames, W. J., List, F. A., Pannala, S., Dehoff, R. R., & Babu, S. S. (2016). The metallurgy and processing science of metal additive manufacturing. *International Materials Reviews, 61*(5), 315–360.

Šaparauskas, J., & Turskis, Z. (2006). Evaluation of construction sustainability by multiple criteria methods. *Technological and Economic Development of Economy, 12*(4), 321–326.

Sartori, I., Napolitano, A., & Voss, K. (2012). Net zero energy buildings: A consistent definition framework. *Energy and Buildings, 48*, 220–232.

Sekhon, B. S. (2014). Nanotechnology in agri-food production: An overview. *Nanotechnology, Science and Applications*, 7, 31–53.

Serpell, A., Kort, J., & Vera, S. (2013). Awareness, actions, drivers and barriers of sustainable construction in Chile. *Technological and Economic Development of Economy, 19*(2), 272–288.

Sethi, P., & Sarangi, S. R. (2017). Internet of Things: Architectures, protocols, and applications. *Journal of Electrical and Computer Engineering, 2017*, 1–25.

Shofoluwe, M. A. (2011). An integrated approach to planning and development of sustainable affordable housing in developing countries. *Sustainable Development & Environmental Protection, 1*(2), 33–42.

Shonnard, D. R., Allen, D. T., Nguyen, N., Austin, S. W., & Hesketh, R. (2003). Green engineering education through a US EPA/Academia collaboration. *Environmental Science & Technology, 37*(23), 5453–5462.

Singh, S., & Agrawal, V. (2021). Critical success factors for new horizons in the supply chain of 3-D printed products – A review. *Materials Today: Proceedings, 44*, 1627–1634.

Singh, T., Shukla, S., Kumar, P., Wahla, V., Bajpai, V. K., & Rather, I. A. (2017). Application of nanotechnology in food science: Perception and overview. *Frontiers in Microbiology, 8*, 1501.

Succar, B., Sher, W., Aranda-Mena, G., & Williams, T. (2007). A proposed framework to investigate building information modelling through knowledge elicitation and visual models. In *Conference Proceedings of the Australasian Universities Building Education Association, Melbourne, July* (pp. 4–5).

Sulzakimin, M., Masrom, M. A. N., Hazli, R., Adaji, A. A., Seow, T. W., & Izwan, A. M. H. (2020). Benefits for public healthcare buildings towards net zero energy buildings (NZEBs): initial reviews. In *IOP Conference Series: Materials Science and Engineering* (Vol. 713, No. 1, p. 012042). IOP Publishing.

Swan, M. (2017). Anticipating the economic benefits of blockchain. *Technology Innovation Management Review, 7*(10), 6–13.

Szydlik, C. (2014). *Identifying and overcoming the barriers to sustainable construction.* Missouri University of Science and Technology.

Thai, H. T., Ngo, T., & Uy, B. (2020). A review on modular construction for high-rise buildings. In *Structures* (Vol. 28, pp. 1265–1290). Elsevier.

Tiwari, S., Singh, R., & Tawaniya, J. (2013). Review on nanotechnology with several aspects. *Int J Res Comput Eng Electron, 2*(3), 1–8.

Tohidi, H. (2011). Review the benefits of using value engineering in information technology project management. *Procedia Computer Science, 3*, 917–924.

United States Environmental Protection Agency. (2016). *Green building.* Retrieved 24 September 2019 from https://archive.epa.gov/greenbuilding/web/html/whybuild.html

United States Green Building Council. (2019). *Leadership in energy and environmental design.* Retrieved 09 August 2019 from https://new.usgbc.org/leed

Upadhyay, A., Mukhuty, S., Kumar, V., & Kazancoglu, Y. (2021). Blockchain technology and the circular economy: Implications for sustainability and social responsibility. *Journal of Cleaner Production, 293*, 126130.

Usmani, Z., Sharma, M., Awasthi, A. K., Sivakumar, N., Lukk, T., Pecoraro, L., Thakur, V. K., Roberts, D., Newbold, J., & Gupta, V. K. (2021). Bioprocessing of waste biomass for sustainable product development and minimizing environmental impact. *Bioresource Technology, 322*, 124548.

Vaigandla, K. K., Karne, R., Siluveru, M., & Kesoju, M. (2023). Review on blockchain technology: Architecture, characteristics, benefits, algorithms. Challenges and Applications. *Mesopotamian Journal of CyberSecurity*, *2023*, 73–85.

Wang, Q., & Su, M. (2020). Integrating blockchain technology into the energy sector – From theory of blockchain to research and application of energy blockchain. *Computer Science Review*, *37*, 100275.

Wells, L., Rismanchi, B., & Aye, L. (2018). A review of net zero energy buildings with reflections on the Australian context. *Energy and Buildings*, *158*, 616–628.

Whang, S., & Kim, S. (2015). Balanced sustainable implementation in the construction industry: The perspective of Korean contractors. *Energy and Buildings*, *96*, 76–85.

Whitmore, A., Agarwal, A., & Da Xu, L. (2015). The Internet of Things – A survey of topics and trends. *Information Systems Frontiers*, *17*, 261–274.

Wong, N. H. (2015). Grand challenges in sustainable design and construction. *Frontiers in Built Environment*, *1*(22), 1–3.

Wong, S. S., & Lau, L. K. (2014). Examining benefits of implementing industrialized building system (IBS) in rural area of developing countries: An exploratory study. In *Proceedings of the 1st International Conference of the CIB Middle East and North Africa Research Network, 14–16 December 2014, Abu Dhabi, United Arab Emirates* (pp. 641–647).

Zabihi, H., Habib, F., & Mirsaeedie, L. (2013). Towards green building: Sustainability approach in building industrialization. *International Journal of Architecture and Urban Development*, *3*(3), 49–56.

Zhou, L., & Lowe, D. J. (2003). Economic principles of sustainable construction. *2nd International Conference on Construction Innovation and Sustainability in the 21st Century* (pp. 660–665).

4 Understanding the Green Building Concept

Introduction

As indicated by Henderson (2012), the green building (GB) of today is not a new concept but a movement initiated by the United States (US) energy crisis of the 1970s, leading to the establishment of the US Environmental Protection Agency (USEPA). Since its early formative years, GB has experienced significant transformation and has since become a mainstream appeal and a 'revolution' sweeping the whole world (Kubba, 2012; Nykamp, 2017). Owing to the global clamour to reduce GHG emissions and climate change issues generally, GB is perceived as a formidable panacea to this menace. To achieve sustainability in the built environment, several countries have embarked on programmes, agendas, missions, and master plans to ensure construction activities and processes are done in an environmentally friendly manner. Many governments are adopting a regulatory approach in their efforts towards addressing the direct and indirect environmental impacts of the construction industry (CI) (Lam et al., 2010). For instance, Singapore aims to achieve the greening of 80% of all buildings in the country by the year 2030 and has rolled out a series of master plans and programmes to this effect (Hwang et al., 2016).

Overview of Green Building

As one of the major concepts of sustainability, GB is a notable term that has gained global acceptability over the years. GB is often used interchangeably with other terms such as sustainable design, green construction, integrated design, sustainable building, sustainable construction (SC), environmentally friendly, green, design of high-performance, and sustainable architecture (Hastings & Wall, 2007; Presley & Meade, 2010; Robichaud & Anantatmula, 2011; Azis et al., 2012; Kubba, 2012; Zabihi & Habib, 2012; Kibert, 2013; AlSanad, 2015). Some researchers regarded GB as a paradigm in reaction to the environmental degradation and crisis facing mankind and the resulting efforts to ensure the efficiency of the built environment through materials, energy, and utilisation (Kubba, 2012). GB, also known as sustainable building (SB), as a concept, strives to meet both environmental and economic goals by leveraging resource-efficient processes throughout the lifecycle of a building. The aim is to address and solve the measurable challenges associated with

DOI: 10.1201/9781003415961-6

the construction and use of traditional buildings (Windapo, 2014). The development of GBs is mainly based on the principles and concepts of SC. However, GBs are believed to be a subset of SC, encompassing the social, economic, and environmental aspects of buildings in the spirit of SD (Kibert, 2004; Hwang et al., 2017).

Definition of Green Building

There are numerous records of GB definitions and differing perspectives of what the term entails. However, there is no globally accepted definition of GB, but a critical review of the existing ones reveals a common thread (Robichaud & Anantatmula, 2011), which is the alignment to sustainability principles. Despite the enormous number of definitions, GBs are regarded as buildings that are resource-efficient, consume less energy, and pose minimal environmental impacts (Hwang & Tan, 2012). According to the definition by Kibert (2016), GB refers to the attributes and qualities of a building structure developed using the methodologies and principles of SC. Anoop et al. (2018) describe GB as construction that minimises the use of non-renewable construction materials and other resources while maximising the use of recycled content and modern efficient engineered materials through efficient engineering design, planning, construction, and effective recycling of construction waste. It is also described as the 'responsible creation and management of a healthy building environment, considering the ecological principles and efficient use of resources' (Kibert, 2013; Carvalho et al., 2019).

Green Building Goals

The literature has established several goals or reasons that encourage the planning, design, construction, and use of GBs. According to Ching and Shapiro (2014), most of the widely recognised GB goals can be regarded as addressing the environmental arm of sustainability and are listed as follows: reduction of air, water, and soil pollution; prevention of unnecessary and irreversible conversion of farmland to non-agricultural uses; protection of clean water sources; reduction in the use of landfills; reduction in the risk of nuclear contamination; reduction of light pollution that can disrupt nocturnal ecosystems; mitigation of global warming through energy conservations; reduction of GHG emissions, and carbon sequestration through biological processes, such as reforestation and wetland restoration; minimisation of environmental impacts resulting from the extraction of natural gas, coal, and oil (including oil spills; the mountaintop removal mining of coal; and the pollution associated with hydraulic fracturing for natural gas); protection of top soil and reduction of the impacts of flooding; and protection of natural habitats and biodiversity, with specific concern for threatened and endangered species. Those regarded as addressing the economic arm include reduction in energy costs, creation of green jobs, improving productivity, improving public relations, and increasing marketing appeal. The goals considered to address improved human health and comfort are improving indoor water quality; improving indoor air quality; reducing noise pollution; increasing thermal comfort; and improving morale. Lastly, the GB goals considered to be political include increasing national competitiveness, reducing strain on electric power grids and risk of power outages, reducing dependence on foreign sources of fuel, and avoiding depletion of non-renewable fuels, such as oil, coal, and natural gas.

The main goal of GB identified by Singh and Sharma (2014) is to minimise the overall negative impact of the CI on both the natural and human environment by using water, energy, and other resources efficiently; reducing pollution, waste, and environmental degradation; and protecting occupant health and improving employee productivity. The study of Shukla et al. (2015) indicated that GBs must embed and achieve the following: conservation of resources and water, improved indoor air quality, building space and green material use, renewable energy generation, energy efficiency, site regeneration, and operational, maintenance and waste recycling. Shi et al. (2014) listed six goals, namely to provide sophisticated and robust technical standards to guide GB developments, to provide a sound legal system with a unified platform to encourage and supervise GB developments, to provide economic incentives to stimulate GB developments, to educate consumers to improve their willingness to accept and patronise GB, to enhance public's awareness and knowledge of GB, and encourage novel delivery and management approaches for GB developments by integrating and engaging various stakeholders. Mehta and Chakraborty (2017) listed the following important to achieving GB: the use of non-toxic and recyclable/recycled materials, minimal disturbance to landscapes and site conditions, reduction of building footprints, measurement and verification plans to ensure water and energy savings, use of onsite renewable energy, and use of energy-efficient equipment for lighting and air conditioning systems. It is therefore important for the GB stakeholders to ensure the aim of building green is attained by incorporating and implementing the fundamental principles earlier stated.

Benefits of Green Building Adoption and Implementation

Literature has established that the adoption and implementation of GB offer numerous benefits to the human and natural environment. The development of GBs has grown significantly over the years, as evidenced by the steadily increasing number of certified/rated buildings globally (Shi et al., 2014). GBs have the potential to bring about water and energy savings, lower operating costs, generate higher rent and sales prices, improve the health of occupants, increase the productivity of occupants, increase the satisfaction of occupants, and provide a conducive environment (Hwang et al., 2017). The USEPA categorised these benefits into three major dimensions in tandem with SD, namely environmental, economic, and social benefits. The perception of GB benefits is aligned with the tripartite pillars of sustainability as further confirmed by Reza et al. (2017).

Environmental Benefits of Green Building

Reduced water streams, enhanced, and protected biodiversity and ecosystems, conserved, and restored natural resources, and improved air and water quality are listed by the USEPA (2016) and Aboulnaga (2013) as the environmental benefits of

GB. Levin (2013) listed the reduced amount of water, energy, and waste generated; improved resource efficiency; and the utilisation of products manufactured from recycled materials and sustainably harvested wood as the environmental benefits of GB. Environmental benefits listed by Say and Wood (2008) are decreased GHG emissions, decreased raw materials use, decreased waste output, decreased fresh-water use, and decreased fuel use.

Economic Benefits of Green Building

Improving occupant productivity, reducing operating costs, optimising life-cycle economic performance, and creating, expanding, and shaping markets for green products and services are listed by the USEPA (2016) and Aboulnaga (2013) as the economic benefits of GB. Levin (2013) highlighted greater employment opportunities; increased desirability for projects because of green features; reduced transportation costs; reduced water and energy bills; earning green rebates and other financial incentives; reduced liability risk from building-associated health issues; and operating cost savings on maintenance, utilities, and replacement costs as environmental benefits of GB. Increased rent value, increased building value, reduced operating costs, and improved employee productivity and satisfaction are identified by the study of Say and Wood (2008) as the economic benefits of GBs.

Social Benefits of Green Building

Improved overall quality of life, heightened aesthetic qualities, minimised strain on local infrastructure, and enhanced occupant comfort and health are listed by the USEPA (2016) and Aboulnaga (2013) as the social benefits of GB. Levin (2013) suggested the following as the social benefits of GBs: improvement of occupants' health and indoor environmental quality (IEQ), the mitigation of harmful health impacts on occupants', the promotion of the use of infill sites and brownfields, the promotion of economic development and liveable communities, and the empowerment of marginalised and underprivileged groups in the community. Considering the significant amount of time human beings spend indoors, GBs are beneficial in the areas of thermal comfort, IEQ, and improved health and productivity (Zuo & Zhao, 2014). Described as health and community benefits, the study of Say and Wood (2008) identified improved overall quality of life, improved thermal comfort, and improved air quality.

Barriers to Green Building Adoption and Implementation

Lack of training and education in sustainable design and construction, procurement issues, lack of awareness on sustainable building, high cost of sustainable building options, lack of demonstration examples, disincentive factors over local material production, lack of professional capabilities, and regulatory barriers are the factors hindering the adoption and implementation of GB (Shafii et al., 2006). The study of Wilson and Rezgui (2013) identified the following as barriers hindering the implementation of GB, namely lack of education and awareness among construction stakeholders, uncertainty of the cost of green solutions and technologies, unclear links between SC principles and current construction standards, lack of access to sustainability knowledge and information, and lack of sharing and exploitation across the CI. Also, the study of Sourani and Sohail (2011) determined the main barriers as resistance to change; lack of a long-term perspective; insufficient research and development; insufficient integration and industry link-up; vagueness of definitions and diversity of interpretations; insufficient time to address sustainability issues; the perceived high cost of addressing sustainability; the separation between capital and operational budget; lack of awareness, understanding, information, commitment, and demand; confusing/insufficient tools, guidance, demonstrations, and best practice; and lack of funding, reluctance to incur a higher capital cost when needed and restrictions on expenditure.

The study of Hankinson and Breytenbach (2012) identified four major categories of GB barriers, namely education and inexperience, cost, client, and materials. A high-cost premium of GB projects, lack of expressed interest from clients and market demand, lack of credible research on GB benefits, high cost of implementing GB practices, and lack of interest and communication among construction stakeholders are the top five barriers identified in the study of Hwang and Tan (2012). The study of Ametepey et al. (2015) identified six factors as barriers, namely financial barriers, political barriers, management/leadership barriers, technical barriers, socio-cultural barriers, and knowledge/awareness barriers. Other identified barriers include misrepresentation or 'greenwashing'; lack of regulatory flexibility; resistance by industry trade unions; need for new suppliers; perceived increased costs; lack of demand; lack of consumer education; process uncertainty; the complexity of certification; and building code issues and interpretation of codes (Marker et al., 2014). The study of Wimala et al. (2016) identified the following barriers to GB adoption and implementation in Indonesia: burdensome implementation; resistance to change; lack of supportive atmospheres; negligence; high cost of GB; inadequate knowledge and information about the GB concept; insufficient supervision from responsible parties; lack of awareness; low availability of green products on the market; lack of building management role in supporting GB movement; risk of investment; high cost of certification; and deficient financial support from the government and credit institutions.

Other challenges to the adoption and implementation of GB include additional costs caused by green construction; incremental time caused by green construction; reduction of structure aesthetics; uncertainty in the performance of green materials and equipment; imperfect green technological specifications; misunderstanding of green technological operations; conflicts in benefits with competitors; regional ambiguities regarding the green concept; restrictions of new green productions and technologies; additional responsibility for construction maintenance; lack of quantitative evaluation tools for green performance; limited availability of green suppliers; dependence on promotion by the government; and lack of support from senior management (Hasan & Zhang, 2016). The study of Saleh and Alalouch (2015) in Oman identified the following

GB barriers: project delays, lack of environmental legislation, cost-effectiveness concerns, lack of knowledge and awareness, and limited availability of green materials and equipment.

Drivers of Green Building Adoption and Implementation

The development of assessment, rating, and labelling systems; establishment of sustainable building and construction champions; development of best practice recognition schemes; creation of building support centres; outreach and awareness raising programmes; education and training programmes; access to information portals; and establishment of research collaboration and exchange programme are suggested by Du Plessis (2005) as ways of promoting the uptake of GB. The study of Serpell et al. (2013) identified awareness, regulations, cost reduction, market differentiation, client demand, corporate image, and suppliers as the drivers that promote GB in Chile. Fiscal measures, legislation, and enabling and support mechanisms such as the establishment of a core environmental assessment tool to provide an integrated medium of measuring sustainability in the CI are suggested by Warnock (2007) as the drivers of GB. Gomes and Da Silva (2005) highlighted four major themes as drivers, namely education and training, research, and data collection for regionalisation of sustainability assessment, financing for acceleration and innovative solutions development, and development and implementation of public and private sector policies.

The study by Davies and Davies (2017) suggested the following as drivers of GB: introducing green construction education into the curriculum of educational institutions, proper awareness of GB, support of government policies, client education, development of green products by suppliers and manufacturers, application of rating tools, and accessibility of information and intricacy of analysis. In addition, set rules and regulations, educational programmes, green design guidelines and standards, and economic incentives are perceived to be the most significant drivers of GB (AlSanad, 2015). Szydlik (2014) identified the following drivers, namely client awareness, building regulations, financial incentives, client demand, taxes, investment, planning policy, and labelling/measurements. According to Windapo (2014), the four main drivers of GB are demand by stakeholders, financial benefits of going green, reduced environmental impact, and the need for corporate/social responsibility. Finally, top management commitment, government regulations, and construction stakeholder pressure are highlighted by Akadiri and Fadiya (2013) as drivers.

Strengthened technology research and communication, introduction and enforcement of codes and regulations (evaluation standards, multi-incentive tools, evaluation systems, and legal management), and enhanced stakeholders' awareness are identified by Li et al. (2014) as ways of promoting GBs. Other factors promoting the uptake of GBs include financial incentives (tax holidays, green loans), public awareness and campaigns, strict enforceable urban land and planning policy, recognition, and certification of sustainable buildings, development of a sustainability checklist by the local authority, improved enforcements by local government, the introduction of mandatory building audits, education, and training focusing on sustainability, promoting green procurement, and investment in GB-related research (Laeeq et al., 2017).

Green Building Collaborative Networks: World Green Building Council

At the forefront of creating awareness and advocating for the uptake and proliferation of GBs globally is the World Green Building Council (WGBC). It is common knowledge that the CI and related activities are the major contributors to GHG emissions that cause climate change worldwide. There is, therefore, an urgent need to reduce environmental degrading acts and embrace sustainable processes, activities, and actions which have the potential to curb this threat. The mission of the WGBC is to create GBs for everyone and everywhere to build a better future with the corresponding goal of accelerating action to ensure all buildings achieve net zero emissions by the year 2050, based on the UN global goals for SD and the Paris Agreement (WGBC, 2016).

The first GB council, the United States Green Building Council (USGBC) was established in 1993. In the year 1998, the WGBC was established (Say & Wood, 2008); in 1999, the founding meeting was held in California, USA, and three years later in 2002, the Council was officially formed with the GB councils of USA, Spain, Mexico, Japan, India, Canada, Brazil, Australia (WGBC, 2016). According to Hydes et al. (2018), the WGBC consists of national GB councils to expedite the radical transformation of the built environment and property industries globally from a conventional to an eco-friendly state. With the presence of GB councils in over 70 countries around the world, the WGBC has grown to become a global network, sensitising, educating, propagating, and advocating for the uptake of GB, GB rating tools, and other processes that help achieve a sustainable and healthier built environment (WGBC, 2016). It is important to note that there are networks of GB councils across Africa, the Americas, Asia-Pacific, Europe, and the Middle East and North Africa regions (MENA). These GB councils are in the membership category of either established, prospective, emerging, or partnership.

Summary

This chapter presented a general overview of the GB concept. To comprehensively understand the fundamentals of GB, this chapter presented the definitions, goals, benefits, barriers, and ways of promoting the adoption and implementation. The chapter also reviews the WGBC, a foremost GB collaborative network with the sole desire to ensure the concept is known, embraced, and implemented in all nations of the world. The next chapter examines and discusses biomimicry (a nature-inspired concept) to further deepen the understanding of sustainable interventions breaking new grounds in the built environment. This is also done to provide the necessary foundation upon which the novel nature-inspired sustainability assessment tool that is conceptualised in this book is birthed.

References

Aboulnaga, M. (2013). Sustainable building for a green and an efficient built environment: New and existing case studies in Dubai. In A. Sayigh (Ed.), *Sustainability, energy and architecture: Case studies in realizing green building* (pp. 131–170). Oxford, UK: Academic Press.

Akadiri, O. P., & Fadiya, O. O. (2013). Empirical analysis of the determinants of environmentally sustainable practices in the UK construction industry. *Construction Innovation, 13*(4), 352–373.

AlSanad, S. (2015). Awareness, drivers, actions, and barriers of sustainable construction in Kuwait. *Procedia Engineering, 118*, 969–983.

Ametepey, O., Aigbavboa, C., & Ansah, K. (2015). Barriers to successful implementation of sustainable construction in the Ghanaian construction industry. *Procedia Manufacturing, 3*, 1682–1689.

Anoop, C. K., Noor, N., James, M., Paul, P. N., & Harikrishnan, R. (2018). Technological and green solutions for rural house construction. *International Research Journal of Engineering and Technology (IRJET), 5*(4), 2291–2293.

Azis, A. A. A., Memon, A. H., Rahman, I. A., Nagapan, S., & Latif, Q. B. A. I. (2012). Challenges faced by construction industry in accomplishing sustainability goals. In *Proceedings of 2012 IEEE Symposium on Business, Engineering and Industrial Applications, 23–26 September 2012, Bandung, Indonesia* (pp. 628–633).

Carvalho, J. P., Bragança, L., & Mateus, R. (2019). Optimising building sustainability assessment using BIM. *Automation in Construction, 102*, 170–182.

Ching, F. D., & Shapiro, I. M. (2014). *Green building illustrated.* Hoboken, New Jersey: John Wiley & Sons.

Davies, O., & Davies, I. (2017). Barriers to implementation of sustainable construction techniques. *MAYFEB Journal of Environmental Science, 2*, 1–9.

Du Plessis, C. (2005). Action for sustainability: Preparing an African plan for sustainable building and construction. *Building Research & Information, 33*(5), 405–415.

Gomes, V., & Da Silva, M. G. (2005). Exploring sustainable construction: Implications from Latin America. *Building Research & Information, 33*(5), 428–440.

Hankinson, M., & Breytenbach, A. (2012). Barriers that impact on the implementation of sustainable design. In *The Proceedings of Cumulus conference, 24–26 May 2012, Helsinki, Finland.*

Hasan, M. S., & Zhang, R. (2016). Critical barriers and challenges in implementation of green construction in China. *International Journal of Current Engineering and Technology, 6*(2), 435–445.

Hastings, R., & Wall, M. (2007). *Sustainable solar housing: Volume 1 – Strategies and solutions.* UK: Earthscan.

Henderson, H. (2012). *Becoming a green building professional: A guide to careers in sustainable architecture, design, engineering, development, and operations.* Hoboken, NJ: John Wiley & Sons.

Hwang, B., & Tan, J. S. (2012). Green building project management: Obstacles and solutions for sustainable development. *Sustainable Development, 20*(5), 335–349.

Hwang, B., Zhu, L., & Ming, J. T. T. (2016). Factors affecting productivity in green building construction projects: The case of Singapore. *Journal of Management in Engineering, 33*(3), 1–12.

Hwang, B., Zhu, L., Wang, Y., & Cheong, X. (2017). Green building construction projects in Singapore: Cost premiums and cost performance. *Project Management Journal, 48*(4), 67–79.

Hydes, K., Richardson, G. R., & Petinelli, G. (2018). World green building council: Supporting the sustainable transformation of the global property market. *World Green Building Council: Supporting the Sustainable Transformation of the Global Property Market*, 49–52.

Kibert, C. J. (2004). Green buildings: An overview of progress. *Journal of Land Use & Environmental Law*, *19*(2), 491–502.

Kibert, C. J. (2013). *Sustainable construction: Green building design and delivery*. John Wiley & Sons.

Kibert, C. J. (2016). *Sustainable construction: Green building design and delivery* (4th ed ed.). New Jersey: John Wiley and Sons Inc.

Kubba, S. (2012). *Handbook of green building design and construction: LEED, BREEAM, and green globes*. USA: Butterworth-Heinemann.

Laeeq, M. Y., Ahmad, S. K., & Altamash, K. (2017). Green building: Concepts and awareness. *International Research Journal of Engineering and Technology*, *4*(7), 3043–3048.

Lam, P. T. I., Chan, E. H. W., Poon, C. S., Chau, C. K., & Chun, K. P. (2010). Factors affecting the implementation of green specifications in construction. *Journal of Environmental Management*, *91*(3), 654–661.

Levin, E. R. (2013). Building communities: The importance of affordable green housing. *National Civic Review*, *102*(2), 36–40.

Li, Y., Yang, L., He, B., & Zhao, D. (2014). Green building in China: Needs great promotion. *Sustainable Cities and Society*, *11*, 1–6.

Marker, A. W., Mason, S. G., & Morrow, P. (2014). Change factors influencing the diffusion and adoption of green building practices. *Performance Improvement Quarterly*, *26*(4), 5–24.

Mehta, S., & Chakraborty, A. (2017). Green buildings and its sustainability: A review. *International Journal of Engineering Technology Science and Research*, *4*(7), 162–167.

Nykamp, H. (2017). A transition to green buildings in Norway. *Environmental Innovation and Societal Transitions*, *24*, 83–93.

Presley, A., & Meade, L. (2010). Benchmarking for sustainability: An application to the sustainable construction industry. *Benchmarking: An International Journal*, *17*(3), 435–451.

Reza, A. K., Islam, M. S., & Shimu, A. A. (2017). Green industry in Bangladesh: An overview. *Environmental Management and Sustainable Development*, *6*(2), 124–143.

Robichaud, L. B., & Anantatmula, V. S. (2011). Greening project management practices for sustainable construction. *Journal of Management in Engineering*, *27*(1), 48–57.

Saleh, M. S., & Alalouch, C. (2015). Towards sustainable construction in Oman: Challenges & opportunities. *Procedia Engineering*, *118*, 177–184.

Say, C., & Wood, A. (2008). Sustainable rating systems around the world. *CTBUH Journal*, *II*(2), 18–29.

Serpell, A , Kort, J., & Vera, S. (2013). Awareness, actions, drivers and barriers of sustainable construction in Chile. *Technological and Economic Development of Economy*, *19*(2), 272–288.

Shafii, F., Ali, Z. A., & Othman, M. Z. (2006). Achieving sustainable construction in the developing countries of Southeast Asia. *6th Asia-Pacific Structural Engineering and Construction Conference (APSEC 2006)*, 29–44.

Shi, Q., Lai, X., Xie, X., & Zuo, J. (2014). Assessment of green building policies– A fuzzy impact matrix approach. *Renewable and Sustainable Energy Reviews*, *36*, 203–211.

Shukla, A., Singh, R., & Shukla, P. (2015). Achieving energy sustainability through green building approach. *Energy sustainability through green energy*, (pp. 147–162), Springer.

Singh, R. R., & Sharma, S. (2014). Green buildings: Opportunities and challenges. In Fulekar, M. H., B. Pathak, & R. K. Kale (Eds.), *Environment and sustainable development* (pp. 177–183). Springer.

Sourani, A., & Sohail, M. (2011). Barriers to addressing sustainable construction in public procurement strategies. *Engineering Sustainability, 164*(4), 229–237.

Szydlik, C. (2014). *Identifying and overcoming the barriers to sustainable construction.* Missouri University of Science and Technology.

United States Environmental Protection Agency. (2016). *Green building.* Retrieved 24 September 2019 from https://archive.epa.gov/greenbuilding/web/html/whybuild. html.

Warnock, A. C. (2007). An overview of integrating instruments to achieve sustainable construction and buildings. *Management of Environmental Quality: An International Journal, 18*(4), 427–441.

Wilson, I. E., & Rezgui, Y. (2013). Barriers to construction industry stakeholders' engagement with sustainability: Toward a shared knowledge experience. *Technological and Economic Development of Economy, 19*(2), 289–309.

Wimala, M., Akmalah, E., & Sururi, M. R. (2016). Breaking through the barriers to green building movement in Indonesia: Insights from building occupants. *Energy Procedia, 100*, 469–474.

Windapo, A. O. (2014). Examination of green building drivers in the South African construction industry: Economics versus ecology. *Sustainability, 6*(9), 6088–6106.

World Green Building Council. (2016). *About us.* Retrieved 23 November 2018 from https:// www.worldgbc.org/our-mission.

Zabihi, H., & Habib, F. (2012). Sustainability in building and construction: Revising definitions and concepts. *International Journal of Emerging Sciences, 2*(4), 570.

Zuo, J., & Zhao, Z. (2014). Green building research – Current status and future agenda: A review. *Renewable and Sustainable Energy Reviews, 30*, 271–281.

5   Biomimicry Paradigm

Nature Inspiration and Emulation

Introduction

Despite the relatively new term 'biomimicry', the application of knowledge and insights gained from the study of natural systems is not a new phenomenon. Since the dawn of time, the natural world has been heavily depended upon for survival, shelter, defence, food, and existence. Strategies, systems, and phenomena in nature were studied and adapted to their advantage by early humans in numerous ways such as manufacturing processes, shelter architectures, armour, sensors and alarm systems, weapons, agriculture, and food production, among others (Murr, 2015). The relationship with nature has always been a symbiotic one to a varying degree, as the depletion of resources in any location drove the early man to another location in search of new and rich resources (Reed, 2004). According to Browning et al. (2014), ancient edifices such as the Egyptian sphinx, Greek temples, and the Neolithic Gobekli Tepe are characterised by stylised leaves and animals' representation for their symbolic and decorative ornamentation. Another example occurred in the Greco-Roman period when architects drew inspiration from trees and plants in the design of columns that structurally resemble tree trunks and the adorning of capitals with acanthus plants (Freitas & Leitão, 2019). As noted by researchers, engineers, and scientists, most of the inventions of the past are inspired by natural models while it is important to recognise that imitating and emulating nature date back to the origins of Westernisation and beyond (Dicks & Blok, 2019). Table 5.1 presents a few examples of early nature-inspired inventions and innovations, coupled with their respective known inventors.

For the over 3.8 billion years that nature has existed, the constituting organisms (flora and fauna) have been able to sustain, survive, evolve, and co-exist among themselves and in the process overcome all challenges. According to El-Zeiny (2012), every organism in nature is distinctive and exclusively adapted to its ecosystem. Waste management, renewable energy generation, efficient use of resources, and a win-win collaboration are a few of the numerous potentials exhibited by the natural world. This has made nature a role model for a synchronised balance and proportion characterised by longevity, efficiency, resource utilisation, and collaboration (Benyus, 1997) as well as an embodiment of novel sources of inspiration. The ability to proffer sustainable solutions to their challenges has enabled the metamorphosis

DOI: 10.1201/9781003415961-7

Table 5.1 Early nature-inspired inventions and innovations

S/N	Summarised title	Inventor	Nature inspiration	Description
1	Proposed design of flying machines (also known as Ornithopter)	Leonardo da Vinci (1452–1519)	Flight of winged animals such as birds and bats and kites	Manned flying machine designed to display Da Vinci's observation, imagination, and enthusiasm powers for the potential of flight. The first successful flight of a manned ornithopter occurred in 1942
2	Tailless glider	Ignaz and Igo Etrich (1904)	Winged seed of Alsomitra macrocarpa, a liana which grows on Pacific islands	The seed with its outgrowth acts as a flying wing and can glide for a significant distance
3	Éole III or Avion III monoplane	Clément Ader (1890)	Bats and bat wings	Tailless monoplane powered by a 20-horsepower steam engine that weighs 51kg and drives a four-bladed propeller
4	London's Crystal Palace at Chatsworth, England	Sir Joseph Paxton (1851)	Huge leaves of the giant Amazonian waterlily	An edifice designed and structured after the huge leaves of the giant Amazonian waterlily
5	Velcro (hook and loop fasteners)	George de Mestral	Seed burs of burdock thistles	Seed burs of burdock thistles hooking onto loops of thread or fur leading to the invention of a hook-and-loop fastener
6	Lightweight tensile structures	Frei Otto	Spider webs	Efficient lightweight tensile structures
7	Camouflage	Hugh Cott	Forest, dragonfly	For concealment and disguise of military personnel during war
8	Thinner and lighter dome structure	Fillipo Brunelleschi (1436)	Eggshells	Thinner and lighter dome for the Florence Cathedral
9	More streamlined ship hulls	Sir George Cayley (1809)	Dolphins	More streamlined ship hulls inspired by studying dolphins

Source: The Biomimicry Institute (2018).

and sustenance of natural organisms over the years, thereby making them a consultative database for efficient and effective solutions and innovations. Also, the inquisitiveness around the natural world has increased with the onward march of science and technology and the need to embrace SD. As urbanisation and civilisation increase, the human environment is faced with several challenges, with nature stepping in to provide useful ideas and solutions. It is therefore upon the premise that nature has sustainably solved similar problems facing the human environment that the concept of biomimicry evolved and became popularised.

Evolution of Biomimicry Terminology

According to Benyus (1997), the term 'biomimicry' (*bi-o-mim-ic-ry*) emanated from the Greek word *bios* (meaning life) and *mimesis* (meaning imitation). The term 'biomimicry' appeared first in the year 1982 as a word, forming part of the topic of Connie L. Merrill's doctoral thesis. The thesis is titled 'Biomimicry of the Dioxygen Active Site in the Copper Proteins Hemocyanin and Cytochrome Oxidase' (Merrill, 1982). After a critical perusal of the thesis, it was discovered that the term was not defined nor portrayed as a field of study as represented by biomimicry proponents. However, Merrill recognised that nature contains 'certain copper metalloproteins which utilise dioxygen as a central theme in their functionality' (Merrill, 1982). Biomimicry as a term became popularised through a book by Janine M. Benyus titled *Biomimicry: Innovation Inspired by Nature* published in the year 1997. Benyus, widely recognised as the originator of this novel field (Goss, 2009) is a biological science writer and author who co-founded the Biomimicry Institute (a non-profit organisation) and Biomimicry 3.8 (a for-profit consultancy). In the words of Paul Hawken of the Natural Capital Institute, 'Janine Benyus is without question the world's most imaginative person in the field of environmental development and restoration. Time spent with Janine is a transmission of hope about what we can learn from and be within nature' (Hargroves & Smith, 2006).

Describing the concept of studying and learning from nature to solve human challenges, related terms such as biomimicry, biomimesis, bioinspiration, bionics, biophilic architecture, biomimetics, biognosis, bioanalogous design, bio-inspired design, biophilia (Benyus, 1997; Vincent et al., 2006; Shu et al., 2011; Gamage & Hyde, 2012; Gruber & Benti, 2013; Kellert, 2014; Ryan et al., 2014; Fewell, 2015; Tempelman et al., 2015; Fayemi et al., 2017; Garcia, 2017; Graeff et al., 2018; Lee & Baek, 2019) are often used interchangeably. As affirmed by Aziz and El Sheriff (2015), no difference exists among these terminologies as they are fundamentally a nature-based concept. However, the only obvious difference can be inferred from the field/sector and the context of their application. All these terms have their root in nature, and they portray a conscious dive into the processes, strategies, forms, and operations of the natural ecosystem in search of sustainable solutions to various issues facing humanity.

The term 'bionics' was coined in the year 1960 by Jack Steele of the United States Air Force at a meeting held at Wright-Patterson Airforce Base in Dayton, Ohio. At this meeting, Steele described bionics as the science of systems that possess functions extracted from the forms, features, and analogues of nature (Vincent et al., 2006). According to Harkness (2004), Dr. Otto Herbert Schmitt, a professor of biophysics, bioengineering, and electrical engineering at the University of Minnesota from the year 1949 to 1983, conceived the term 'biomimetics'. Biomimetics was first used as part of the paper title presented by Schmitt in the year 1969 at the 3rd International Biophysics Congress held in Boston. Schmitt's doctoral research strove to create a gadget that thoroughly emulated the electrical activity of a nerve. Also, in the year 1974, biomimetics as a word first appeared in Webster's Dictionary. It was described as the study of forms, structures, and functions of biologically produced materials (such as silks or enzymes) and biological processes and mechanisms (such as photosynthesis and protein synthesis) for synthesising similar materials by man-made procedures which emulate nature (Harkness, 2004; Vincent et al., 2006).

Definition of Biomimicry

Despite being a novel field of study, there are several definitions of biomimicry based on the understanding of different researchers in the domain. However, all these definitions are aligned to the natural world in parts or its entirety. Biomimicry is defined as the evaluation of natural elements, systems, procedures, and strategies with the ability and capacity to solve human problems (ElDin et al., 2016). According to Benyus (1997), biomimicry is the quest of humanity to explore amazing nature's geniuses (such as self-assembly, self-cleaning, and photosynthesis) and then copy these strategies to solve their challenges. Biomimicry is further described by Singh and Nayyar (2015) and Nkandu and Alibaba (2018) as the process of initiating sustainable design solutions through the study and purposeful emulation of nature's forms, processes, and ecosystems. In biomimicry, sustainable solutions to human challenges are birthed through the emulation of nature's survival attributes (Badarnah & Kadri, 2015). As informed by Ryan et al. (2014), the holistic application of biomimicry explores nature in a restorative and equally inspirational manner without disturbing the functions of its ecosystem to which it is integral.

Biomimicry is founded on the premise that nature has sustainably solved challenges like most problems ravaging the human environment (Tempelman et al., 2015). The natural world also provides a plethora of capabilities and monumental potentials to learn from in achieving different sustainability goals (Eadie & Ghosh, 2011). According to Duane Elgin, the author of *Voluntary Simplicity*, 'biomimicry fires the imagination with the exciting possibility of taking the best designs from nature's storehouses of the invention and applying them to the challenges of building a sustainable and creative future'. Biomimicry has become a disruptive path to achieving sustainable developments through nature emulation at all stages (from scoping to creation and evaluation) of the design process (Kenny et al., 2012). Biomimicry is now spearheading technological and scientific research owing to its potential to inspire novel insights for evolving sustainable and innovative solutions to human challenges (Zhang et al., 2015). In the quest for sustainable methodologies, practices, and solutions, biomimics, professionals, and other sustainability proponents are now consulting and concentrating on nature to learn from its over 3.8 billion years of evolution (Benyus, 2011).

Biomimicry Dimensions for Optimising Sustainability

The natural world remains the core and underlying concept of biomimicry. It is therefore important for the human relationship with nature to be from the perspective of a 'model, measure, and mentor' (Benyus, 1997; Arnarson, 2011). Therefore, for biomimicry to be able to solve human challenges and achieve its overarching goal of sustainability, the relationship with nature should be oriented towards these three dimensions.

Biomimicry as a Model

With nature as a model, the forms, processes, strategies, and systems exhibited by natural organisms are studied to inspire solutions and methods of tackling human challenges sustainably (ElDin et al., 2016). An example is a solar cell inspired by a leaf, bearing in mind that biomimicry studies nature's models and draws inspiration from their processes and designs to solve human challenges (Benyus, 1997). This perception of nature has the potential to birth numerous outstanding innovations inspired by the amazing systems and processes in nature.

Biomimicry as a Measure

Nature as a measure pertains to the evaluation of human designs, solutions, and innovations against the principles of nature (biomimicry principles) to ascertain their level of rightness and compliance with sustainability yardsticks (Pronk et al., 2008; ElDin et al., 2016). Owing to its over 3.8 billion years of existence, nature has mastered what is effective, efficient, lasting, and appropriate. Hence, biomimicry employs an ecological standard (nature principles/biomimicry principles) to determine the maintainability of our advancements and the rightness of our innovations (Benyus, 1997; Ferwati et al., 2019). According to Goss (2009) and Nkandu and Alibaba (2018), certain questions must be asked, and positive answers sought to ascertain the conformity of biomimicry applications to sustainability, such as Is the design locally attuned and responsive? Is it informed by local inhabitants of all species? Is it resourceful and connected to local feedback loops? Does the design integrate cycles? Does it adapt to seasons, reuse materials, and maintain itself through turnover? Is the design resilient? Can it withstand disturbance while maintaining function? Does it heal after a disturbance? Does the design optimise rather than maximise? Does it fit form to function? Does it reuse materials or use recycled materials? Does the design use benign manufacturing? Is the reaction done at standard pressure and temperature? Does the design leverage its interdependence in the system? Does it enhance the system's capacity to support life long-term? Is its success based on whether it contributes to the continuity of life?

Biomimicry as a Mentor

By looking to nature as a mentor, humans' attitude, and idea of exploiting the natural world and its rich resources are changed to a rapt and focused attention on what can be learnt and emulated (Arnarson, 2011). Biomimicry is described as a novel way of valuing and viewing nature while introducing an era of learning from nature and not exploiting nature (Benyus, 1997; Pronk et al., 2008). Humans must develop a purposive goal of learning from nature and its geniuses while also cultivating a mutual and ethical relationship and consciousness that nature is a constituent of human existence. Biomimicry can be employed to create novel and sustainable innovations and inventions (Zari, 2007), and can also serve as a medium for optimising the sustainability performance of existing creations (Okuyucu, 2015). Attesting to the vast deposit of potential in nature, Apple co-founder, Steve Jobs, said: '…I think the biggest innovations of the 21st century will be at the intersection of biology and technology. A new era is beginning…' (Stafford, 2011).

Characteristics of Biomimicry Based on Application

Considering its application, different opinions have been projected regarding the levels of biomimicry application. As a result, biomimicry is classified into two broad views based on its level of application (form, process, and ecosystems). Table 5.2 presents the constituting attributes of the forms, processes, and ecosystem relationships in nature. As further explained by El-Zeiny (2012), the form, process, and ecosystem levels of biomimicry application can be regarded in tandem with the following biological order, namely organism features (features of the organism itself such as the form, shape, figure, and colour); organism-community relationship (organism's interrelation with its community of similar or other organisms); and organism-environment relationship (organism's interrelation with its environment and the biome at large).

Views of Biomimicry Application for Sustainability

The point from which the natural environment is perceived will significantly determine the level and depth of its application. According to Volstad and Boks (2008), biomimicry applications can either be viewed to be reductive (shallow biomimicry) or holistic (deep biomimicry) as depicted in Figure 5.1. Application of biomimicry principles from any of these views can either incorporate a pillar of sustainability or all of it holistically. However, biomimicry proponents suggest that to achieve all-encompassing sustainability, which is the overarching goal of biomimicry, it must be incorporated and applied in its deepest form.

Views of Biomimicry Application: Form

Biomimicry application based on the *form* of natural organisms falls within the reductive view as shown in Figure 5.1. In this category, biomimicry is perceived and applied based on the performance and sustainable attributes of the form, shape, look, or visible features of an organism in nature. Also known as the traditional or basic kind of biomimicry, this category focuses solely on the emulation of the functional features of an organism with no clear-cut goal of achieving sustainability (Reap et al., 2005; Volstad & Boks, 2008). This biomimicry level is known as the

Table 5.2 Characteristics of biomimicry levels

S/N	Application levels	Description
1	Organisms Forms/ Features (features of the organisms itself)	Formal attributes include shape, colour, transparency, rhythm, volumetric treatment
		Construction materials and process
		Structure, stability, and gravity resistance
		Function and behaviour
		Mutation, growth, and lifecycle
		Organisation and hierarchy of parts and systems
		Motion and aerodynamics
		Morphology, anatomy, modularity, and patterns
		Self-assembly
		Portability and mobility
		Healing, recovery, survival, and maintenance
		Homeostasis that balances internal systems while external forces change
		Systems which include organ, digestive, circulatory, respiratory, skeletal, muscular, nervous, excretory, sensory, and locomotive systems
2	Organisms Strategies/ Techniques/Processes (the relationship of organisms with its community of similar or other organisms that it may interact/ deal with)	Survival techniques
		Transgenerational knowledge transfer and training
		Interaction with other creatures
		Hierarchy of community members
		Group management and coordination
		Collaboration and teamwork
		Self-protection
		Sensing, responding, and interaction
		Communication
		Risk management
3	Organisms Ecosystem (the fitting and relationship of an organism in its biome or environment)	Adjustment to change
		The contextual fit
		Response to climate by cooling, heating, and ventilation solutions
		Adaptation to ecosystems e.g. adjustments to various light or sound levels, shading, and self-illumination
		Response to context e.g., camouflaging, self-protection, and self-cleaning
		Waste management
		Shelter building
		Input/output process cycling
		Limited resource management such as adaptation to the lack of water, light, or food

Source: El-Zeiny (2012).

most straightforward as it entails the emulation of a part or whole form/structure of an organism (Zari & Storey, 2007; Kenny et al., 2012). Examples include the emulation of the kingfisher's beak as the innovative idea behind the redesign of Japan's famous bullet train, thereby solving its noise issues. Another example is the lightweight but structurally strong building panels inspired by the support ribs and shapes of the giant leaves of the Amazon water lily (Attenborough & Graham,

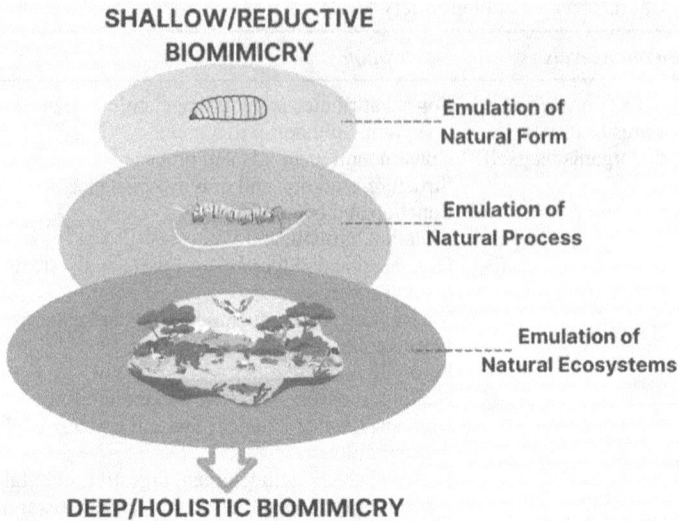

Figure 5.1 Biomimicry views based on application

1995; Kennedy et al., 2015). Innovations in this category will solve a problem; however, the whole system cannot be labelled as fully biomimetic (sustainable) because it does not perform at the holistic biomimicry level.

Views of Biomimicry Application: Process

Biomimicry application based on the *process* utilised by natural organisms falls in between the shallow and the deep categories as shown in Figure 5.1. In this category, the application of biomimicry is based on the emulation of not just the forms, shapes, or appearances of organisms but also the strategies and processes that allow them to thrive well in their ecosystem. It is a deeper level compared to the previous as it studies and emulates the behaviours and series of operations in organisms (recycling, building, manufacturing, among others). While assembling structures at ambient temperature and pressure utilising non-toxic chemistry when compared to human methods (Kennedy et al., 2015), nature is believed to be energy efficient, less wasteful, and non-polluting. A few of these processes are those that allow organisms to be multifunctional, adaptive, self-assembly, self-cleaning, self-healing, self-repairing, superhydrophobic, aerody-namic, antireflective, highly adhesive, or thermo-resistant, amongst numerous others (Bhushan, 2009; Eadie & Ghosh, 2011). An example of applications in this category is the creation of algorithms by computer scientists based on the movement coordination of flocking birds and swarming bees. Another example is the attempt in the field of green chemistry by emulating the benign process of how owl feathers self-assemble at normal temperature without high pressure or toxins (Benyus, 2011).

Views of Biomimicry Application: Ecosystem

Biomimicry application based on the *ecosystem* of natural organisms falls within the deep or holistic biomimicry as shown in Figure 5.1. This level relates to the performance and behaviour of natural organisms in tandem with the ecosystem. This category aims to achieve the goals of sustainability by considering the overall organism-ecosystem relationship where organisms thrive without causing harm to their ecosystem (Volstad & Boks, 2012). In nature, no organism operates or exists in isolation (Kenny et al., 2012), as each one is part of a biome that is part of a larger biosphere (Kennedy et al., 2015). Along similar lines, Benyus (2011) pointed out the gracefully nested owl feather which is part of an owl that is part of a biome that is an integral part of a sustaining biosphere. It is therefore imperative for any innovation to be part of a larger economy that seeks to sustain and restore rather than deplete the environment. Natural organisms survive and sustain themselves through their ability to operate and perform in tandem with their ecosystem. As affirmed by McDonough and Braungart (2010) and Kennedy et al. (2015), every organism's continued flourishing is solely dependent on the health of its ecosystem or biosphere. It is when biomimicry considers the forms, processes, and ecosystem relationships of organisms that it attains the deep/holistic category where it achieves its overarching goal of sustainability. It is only at this level that biomimicry assures and proffers truly sustainable solutions to human challenges as it entails the application of the three levels (natural forms, natural process, and natural ecosystem) in a responsible manner. Hence, biomimicry proponents advocate for the practice and application of biomimicry at the deep/holistic level (incorporating the forms, processes, and ecosystem of organisms) in achieving sustainable solutions to human issues.

Biomimicry Application: Design Strategies

As identified by Vincent et al. (2006), the lack of a well-defined approach is a major barrier to the applicability of biomimicry, hence limiting its potential and success in providing innovative solutions in the built environment and other sectors. Biomimicry approaches or design strategies are the simplified stages that aid the successful application of biomimicry. According to Graeff et al. (2019), the processes of these approaches are based on the extraction of biological knowledge into comprehensive strategies for solving various problems while avoiding the issues related to utilising biological data in raw form. According to Niewiarowski and Paige (2011) and Pandremenos et al. (2012), biomimicry approaches are the step-by-step and simplified avenues through which nature-inspired knowledge is abstracted into various disciplines to proffer sustainable solutions to human challenges. These approaches/strategies are formalised processes and methods through which the application of biomimicry is systemised to solve different human challenges. The studies of Fayemi et al. (2017) and Graeff et al. (2018) identified eight unified steps/processes that are abstracted from nature to apply biomimicry successfully, namely problem analysis, abstract technical problem, transpose to biology, identify potential biological models, select biological models of interest, abstract biological strategies, transpose to technology, and implement and test in the initial context. Along similar lines, the eight definite steps identified by The Biomimicry Institute (2013a) are: define (context), identify (function), integrate (life's principles), discover (natural models), abstract (biological strategies), brainstorm (bio-inspired ideas), emulate (design principles), and measure (using life's principles). These eight steps are further categorised into four classes, namely scoping (containing *define*, *identify*, and *integrate*), discovering (containing *discover* and *abstract*), creating (containing *brainstorm* and *emulate*), and emulating (containing *measure*).

The first three steps, namely define, identify, and integrate, are all embedded in the *scoping* category. This category is the stage where the problem and design questions are formulated and thereafter translated into biological terrain. This stage is important in the process of applying biomimicry successfully. Albert Einstein, the renowned German physicist said, 'If I were given one hour to save the world, I would spend fifty-nine minutes defining the problem and one minute solving it', hence establishing the significance of this category in the biomimicry process. In this category, the problem is well-defined while also identifying the verbs (which describe functions such as distribute, move, repel, breakdown, conserve, and transport, amongst others) that directly define the problem. To fully understand these function-describing verbs, the biomimicry taxonomy is conceptualised by the Biomimicry Institute. As indicated by The Biomimicry Institute (2013b), the biomimicry taxonomy, developed by the Biomimicry Institute, is a visually represented system of classification to organise biological contents on the Asknature database. The biomimicry taxonomy categorises the numerous ways through which natural systems and organisms meet their functional challenges. Afterwards, the described functions are evaluated against biomimicry principles (inferred from life's/nature's principles) and then the problem is 'biologised' (integrated into biology) to properly understand where it fits in nature. At this stage, biomimicry proponents support a robust multidisciplinary collaboration where biologists and nature conservationists are invited to the design table for their input, depending on how complicated the problem to be tackled is. To achieve the overarching goal of biomimicry which is sustainability (holistic), both interdisciplinary and multidisciplinary collaboration are necessary and must therefore be employed.

The *discovering* category consists of the discover and abstract steps as earlier highlighted. This category entails the discovery of natural models that best fit the formulated design problems and the abstraction of biological strategies exhibited by the discovered natural model. To be successful in this category, it is suggested that questions such as: How does nature discharge that function? How does nature not discharge that function? and How does nature discharge that function here? in these unique conditions are asked after identifying the desired function to be performed. Afterwards, natural organisms exhibiting the relevant and intended strategies are pinpointed so that they can be emulated.

The *creating* category consists of the 'brainstorm' and 'emulate' steps. It is in this category that professionals and other stakeholders who are involved in a task extensively brainstorm nature-inspired ideas and subsequently translate the ideas from the biological terrain to technical solutions. The anticipated innovation and solutions to identified problems are developed in this category. The *evaluating* category is the last of the four categories and it contains the measure step. This category remains a significant part of the process of applying biomimicry as it ensures that the resultant outcome, solutions, and innovations are truly sustainable. Outcomes are evaluated against biomimicry principles (life's principles) to ascertain the compliance level in achieving the goals of sustainability. As a result, solutions and innovations can be said to be truly biomimetic (sustainable) when they have satisfied all the requirements as outlined in biomimicry principles. In summary, this category entails determining the sustainability attainment of innovative ideas and solutions by measuring them using life's principles (biomimicry principles).

Biomimicry Application: Approaches

Two key approaches can be employed in the application of biomimicry although they are represented through different terminologies by authors and researchers globally. They are the solution-based and the problem-based biomimicry approaches which are the adopted terminology in this research study. These two approaches or design strategies contain the eight steps of defining (context), identifying (function), integrating (life's principles), discovering (natural models), abstracting (biological strategies), brainstorming (bio-inspired ideas), emulating (design principles), and measuring (using life's principles) as postulated by The Biomimicry Institute (2013a). They are further categorised into four classes, namely scoping (containing *define, identify*, and *integrate*), discovering (containing *discover* and *abstract*), creating (containing *brainstorm* and *emulate*), and emulating (containing *measure*). However, what differentiates the solution-based approach, and the problem-based approach is the indefinite and definite path and strategies (procedural transitions) they follow in their respective application processes. Despite the difference in the two approaches (solution-based and problem-based), a similar trend is observed in both which is the transfer of knowledge and innovative ideas from the biological domain to the technological state (Badarnah & Kadri, 2015).

Solution-based Biomimicry Application Approach

The solution-based approach, according to El-Zeiny (2012) and Badarnah and Kadri (2015) is also referred to as the solution-driven biologically-inspired approach (Helms et al., 2009), solution-driven approach (Shu et al., 2011), solution-driven biologically-inspired design process (Vattam et al., 2007), bottom-top approach (Knippers & Speck, 2012), solution-to-problem approach (Pandremenos et al., 2012; Taylor Buck, 2015), bottom-up process (Speck & Speck, 2008), biology to design (The Biomimicry Institute, 2013a), and biology influencing design approach (Zari, 2007). In the solution-based approach, a specific function, behaviour, or attribute in an organism or ecosystem is identified and then emulated to innovate solutions and outstanding human designs (Zari, 2007). To successfully utilise this approach, Fayemi et al. (2017) suggested that the biological system exhibiting a function of interest must be well understood to extract underlying principles to be emulated in the technological outcomes. Further supported by El-Zeiny (2012), the successful application of this approach depends largely on the professionals and stakeholders having the requisite knowledge of the ecological or biological research on the organism of interest rather than on determined human challenges.

Authors and researchers presented the processes to be followed in this approach in different ways. However, a detailed comprehension of these steps or processes shows that they all represent the same intention and result. According to Vattam et al. (2007), biological solution identification, problem definition and principle extraction, humanisation of the problem-solution pair, problem search, and principle application are the respective steps to be followed in this approach. Building upon this study, El-Zeiny (2012) listed the steps followed in this approach as follows: biological solution identification, defining the biological solution, principle extraction, reframing the solution, problem search, problem definition, and principle application. To simplify these steps for easy understanding and application of the solution-based biomimicry approach, The Biomimicry Institute (2013a) presented eight steps to be followed in a definite order and path. The steps are: discover natural models, abstract biological strategies, identify function, define context, brainstorm bio-inspired ideas, integrate life's principles, emulate design principles, and measure using life's principles.

To apply this approach as presented by The Biomimicry Institute (2013a), the unique strategies possessed and exhibited by organisms and ecosystems are identified and studied (discover natural models). This step is followed by unravelling and determining the techniques behind the unique strategies exhibited by these organisms and ecosystems before translating them into a design principle (abstract biological strategies). It is important to remove all biological references in this step. By using the design principle and strategies exhibited in organisms and ecosystems as a guide, the functional need to be met or the purpose to be fulfilled is defined in this step (identify function). The next step is specifying the actual circumstances where the identified function is required by asking questions such as 'Who or what needs to do what this organism or ecosystem is doing?' (define context). The next step entails thinking of ideas and ways of combining the context, design principle, and function in solving a challenge or design problem (brainstorming bio-inspired ideas). While brainstorming the ideas considered, this step informs on incorporating biomimicry principles (life's principles) into the solution to ensure they are sustainable (integrate life's principles). The next step entails considering the aspects of scale, and the possibility of going beyond emulation at form level to process and ecosystem levels. This is done after selecting the best ideas from the brainstorming step and developing a design concept (by emulating design principles). In the last step, the resultant design or solution generated is evaluated against biomimicry principles (using life's principles) to ascertain the level of sustainability compliance (by measuring using life's principles). Figure 5.2 shows the application path of the solution-based biomimicry approach.

A notable example of the innovative solutions borne out of the solution-based approach to biomimicry application is the self-cleaning smart paint called

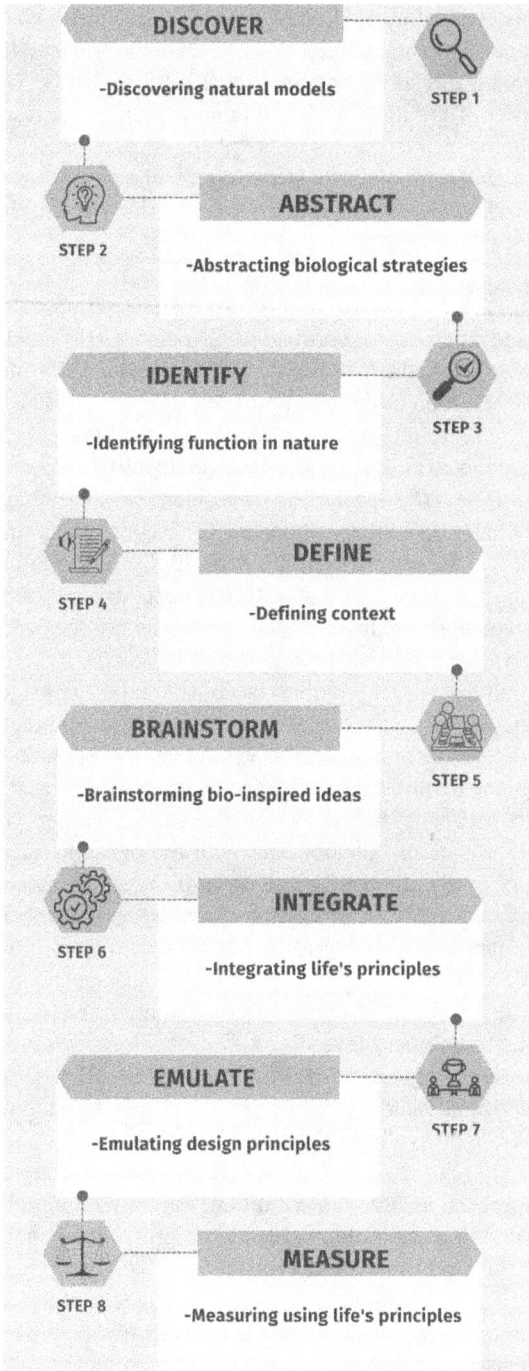

Figure 5.2 Application path of the solution-based biomimicry approach

Lotus Paint. It was inspired by the Lotus leaf, the surface of which contains packed tiny bumps or ridges preventing water drops from sticking to the leaves but rather rolling off, removing dirt along the surface (El-Zeiny, 2012). The hydrophobic nature of the leaves of the lotus plant (Nelumbo Nucifera) is responsible for the amazing self-cleaning characteristics of the plant (Alexandridis, 2016). In this approach, the Lotus leaf is the point of departure for the design, resulting in a self-cleaning paint which is highly useful in the construction industry (CI).

Problem-based Biomimicry Application Approach

The problem-based approach according to El-Zeiny (2012) and Bardanah and Kadri (2015), is also referred to as a problem-driven biologically-inspired approach (Helms et al., 2009), problem-driven approach (Shu et al., 2011), problem-driven biologically-inspired design process (Vattam et al., 2007), design-to-biology approach (Pandremenos et al., 2012), top-down approach (Knippers & Speck, 2012), top-down process (Speck & Speck, 2008), problem-to-solution approach (Taylor Buck, 2015), challenge to biology approach (The Biomimicry Institute, 2013a), and design looking to biology approach (Zari, 2007). In the problem-based biomimicry approach, the living world of nature is depended on for innovative solutions by first recognising the problems the biologists are required to match with organisms that have tackled related situations (Zari, 2007). According to Gamage and Hyde (2012), this approach involves establishing the challenge to be resolved and thereafter studying and conceptualising the strategies and methodologies utilised by natural organisms or ecosystems in settling corresponding issues. Firstly, human problems are identified and defined before pairing such problems with natural organisms and ecosystems that have tackled a similar case. Zari (2007) reiterated the importance of collaborating with biologists to successfully match identified challenges with organisms that have resolved similar issues. The problem-based biomimicry approach is widely practised by engineers, innovators, and designers (El Ahmar, 2011), owing to its potential to transform the architecture, engineering, and construction (AEC) industry into a sustainable one (McDonough & Braungart, 2010; El-Zeiny, 2012) by offering solutions to identified challenges at the deep/holistic biomimicry level.

In this approach, a progression of different steps is followed which are found to be dynamic and nonlinear, as the output from the next stages influences the previous ones thereby providing refinement loops and iterative feedback (Helms et al., 2009). The study of El-Zeiny (2012) identified the six steps followed in the solution-based approach as problem definition, reframing the problem, biological solution search, defining the biological solution, principle extraction, and principle application. The study of Graeff et al. (2019) identified eight steps adapted from the study of Fayemi et al. (2017). They are problem analysis, abstraction of the technical problems, transposition to biology, identification of biological models, selection of models of interest, abstraction of biological strategies, transposition to technology, and implementation in the initial context and test. Vattam et al. (2007) identified seven steps, namely problem definition, problem decomposition, 'biologise' the problem, biological search, define the problem, principle extraction, and principle application. However, in a bid to propagate the application of bio-mimicry across all disciplines, The Biomimicry Institute (2013a) postulated eight steps in a simplified form in a definite order and path. The steps are: define context, identify function, integrate life's principles, discover natural models, abstract biological strategies, brainstorm bio-inspired ideas, emulate design principles, and measure using life's principles.

Applying this approach as presented by The Biomimicry Institute (2013a), the challenge to be tackled and its operating conditions are specified (define context). The key function(s) the design is intended to perform is determined by asking what it needs to do (identify function). The next step is committing to incorporate life's principles (biomimicry principles) into the design requirements (integrate life's principles). This is followed by finding organisms or ecosystems that have evolved strategies to solve the function(s) needed (discover natural models). The mechanisms behind each organism's strategy are determined and translated into a design principle while remembering to remove biological references (abstract biological strategies). The following step is to celebrate several ideas on how to apply the design principles to solve the challenge at hand (brainstorm bio-inspired ideas). The next step considers the aspects of scale after selecting the best ideas from the brainstorming step and developing a design concept with the possibility of going beyond emulation at the form level to process and ecosystem levels (by emulating design principles). In the last step, the solution generated is evaluated against biomimicry principles (using life's principles) to ascertain the level of adherence to sustainability criteria (by measuring using life's principles). Figure 5.3 shows the application path of the problem-based biomimicry approach.

An example of innovative solutions as a result of applying the problem-based bi-omimicry approach is the redesign of Japan's bullet train known as the Shinkansen. The idea behind the redesign is inspired by the long narrow beak of the kingfisher which helps them to transit quickly from a low-drag medium (air) to a high-drag medium (water). The problem to be solved with the Shinkansen high-speed bullet train is the booming loud sound it makes especially when exiting a typical train tunnel. With the shape of the kingfisher's beak, steadily increasing in diameter from the tip to the head, there is a reduction in impact as the kingfisher finds its way into the water, allowing water to flow past the beak rather than being dragged by it (The Biomimicry Institute, 2018). By emulating this outstanding strategy exhibited by the Kingfisher, the redesign of the Shinkansen high-speed bullet train resulted in a more streamlined train that uses less energy and travels faster and more quietly.

Biomimicry Principles

Biomimicry principles, also referred to as Life's/Nature's principles, are specific sets of values exhibited by natural organisms to live, thrive, and survive in the over 3.8 billion years of their existence, enabling each organism to be uniquely and fully adapted to its ecosystem. According to Pólit and David (2014), these values are abstracted biological strategies, processes, forms, and actions found in natural

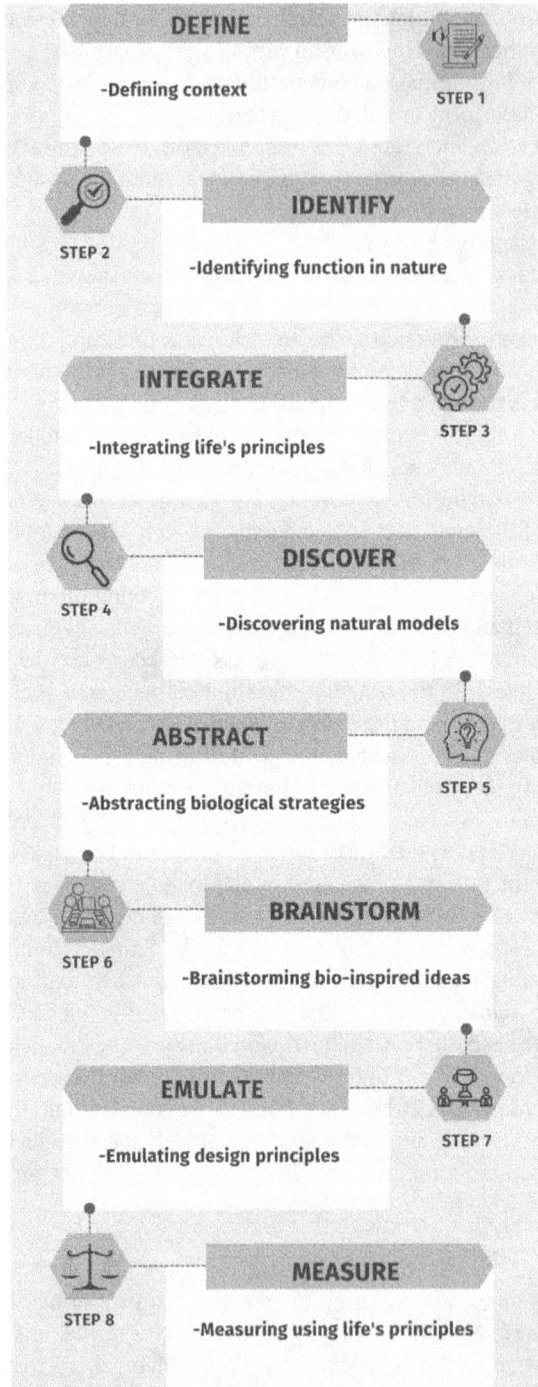

Figure 5.3 Application path of the problem-based biomimicry approach

organisms, most of which are obvious, ensuring life successfully regenerates itself. Benyus (1997) described them as the canon of principles, laws, and strategies divined from nature's notebooks. The nine principles of nature are the following:

- Nature runs on sunlight;
- Nature uses only the energy it needs;
- Nature fits form to function;
- Nature recycles everything;
- Nature rewards cooperation;
- Nature banks on diversity;
- Nature demands local expertise;
- Nature curbs excesses from within; and
- Nature taps the power of limits (Benyus, 1997; Neill, 2018).

Another compilation of nature principles is found in the book titled *Patterns: Sixteen Things You Should Know About Life* by Mahlon Hoagland and Bert Dodson. From their expanded viewpoint, life (nature) is perceived in terms of patterns and rules upon which it builds, organises, recycles, and recreates itself. The 16 principles/patterns of life identified and claimed not to be indefinite by Hoagland and Dodson (1999) and Scheffer (2016) are the following:

- Life builds from the bottom up;
- Life assembles itself into chains;
- Life needs an inside and an outside;
- Life uses a few themes to generate many variations;
- Life organises with information;
- Life encourages variety by reshuffling information;
- Life creates with mistakes;
- Life occurs in water;
- Life runs on sugar;
- Life works in cycles;
- Life recycles everything it uses;
- Life maintains itself by turnover;
- Life tends to optimise rather than maximise;
- Life is opportunistic;
- Life competes within a cooperative framework; and
- Life is interconnected and interdependent.

According to Ferwati et al. (2019), a collaboration between Dayna Baumeister of the Biomimicry Guild and Taryn Mead, the developer of the Fully Integrated Thinking (FIT) project at HOK documented 15 sustainability potentials found in nature. Table 5.3 presents the FIT system that incorporates the three dimensions of sustainability (social, economic, and environmental).

Biomimicry principles emanate from life's principles and are employed to drive and assess the appropriateness and sustainability adherence of biomimicry

Table 5.3 The fully integrated thinking matrix for sustainability

Sustainability dimension	Realm	Functions	Goals
Environmental	Eco-structure	Provide ecosystem services	Maintains and fosters the health and integrity of the native physical and ecological landscape
	Water	Manage water	Protects and enhances water quantity and quality
	Atmosphere	Protects air	Protects and enhances air quality
	Materials	Manage materials	Fosters closed material loops and eliminates waste
	Energy	Provide energy	Provides safe, clean, abundant, reliable, consistent, free energy for all inhabitants in perpetuity
	Food	Provide sustenance	Provides safe, clean, abundant, reliable, consistent, free sustenance for all inhabitants in perpetuity
Social	Community	Foster community	Fosters integrated, connected community identity for all inhabitants
	Culture	Support cultural exchange	Reflects a vibrant exchange of historical and modern identity, food, art, music, and science rooted in place
	Health	Promote health	Ensures health and well-being for all citizens and universal access to quality healthcare
	Education	Provide education	Fosters world-class, life-long learning opportunities for all citizens
	Governance		Maintains responsive accountable stewardship
	Transport	Provide comfort	Provides congestion-and-pollution-free mobility
	Shelter	Provide mobility	Protects inhabitants comfortably from biotic and abiotic factors
		Provide comfort	
Economic	Commerce	Foster commerce	Fosters the balanced exchange of goods and services
	Value	Provide value	Sustains value for investors

Source: Mead (2011) and Ferwati et al. (2019).

applications (Ásgeirsdóttir, 2013). Biomimicry principles (nature/life's principles) can serve as a tool or a process to support the integrated and holistic efforts towards achieving the goals and objectives of the three dimensions of SD. They are important checklists to be adhered to in ensuring the application of biomimicry resulting in sustainable outcomes, either through the problem-based or the solution-based approach methodology. The development of the principles is inspired by the characteristics of the sustainable solutions proffered by natural organisms to their problems. As postulated by the Biomimicry Institute, biomimicry principles are the expanded and detailed version of the principles of nature (life) by Janine Benyus (Biomimicry 3.8, 2017). These principles indicated that for sustainability to be achieved, the system must evolve to survive, be locally attuned and responsive, integrate development with growth, be a resource (material and energy) efficient, use life-friendly chemistry, and adapt to changing conditions.

Evolve to Survive

This is the continuous incorporation and embodying of information to ensure enduring performance. It consists of three sub-principles, namely replicating strategies that work (repeat successful approaches); integrating the unexpected (incorporate mistakes in ways that can lead to new forms and functions); and information reshuffling (exchange and alter information to create new options).

Resource Efficiency: Material and Energy

This is skillfully and conservatively taking advantage of resources and opportunities. It consists of four sub-principles, namely using multifunctional design (meet multiple needs with one elegant solution); using low energy processes (minimise energy consumption by reducing requisite temperatures, pressures, and/or time for reactions); recycling all materials (keep all materials in a closed loop); and fitting form to function (select shape or pattern based on need).

Adapt to Changing Conditions

This is appropriately responding to dynamic contexts. It consists of three sub-principles, namely maintaining integrity through self-renewal (persist by constantly adding energy and matter to heal and improve the system); embodying resilience through variation, redundancy, and decentralisation (maintain function following disturbance by incorporating a variety of duplicate forms, processes, or systems that are not located exclusively together); and incorporating diversity (include multiple forms, processes, or systems to meet a functional need).

Integrate Development with Growth

This entails optimally investing and engaging in strategies that promote both development and growth. It also consists of three sub-principles, namely combining modular and nested components (fit multiple units within each other progressively from simple to complex); building from the bottom up (assemble components one unit at a time); and self-organising (create conditions to allow components to inter-act in concert to move towards an enriched system).

Be Locally Attuned and Responsive

This entails fitting into and integrating with the surrounding environment. It con-sists of four sub-principles, namely using readily available materials and energy (build with abundant, accessible materials while harnessing freely available energy); cultivating cooperative relationships (find value through win-win interac-tions); leveraging cyclic processes (take advantage of phenomena that repeat them-selves); and using feedback loops (engage in cyclic information flows to modify a reaction appropriately).

Use Life-friendly Chemistry

This entails the use of chemistry that supports life processes. It also consists of three subprinciples, namely building selectively with a small subset of elements (assembling relatively few elements in elegant ways); breaking down products into benign constituents (using chemistry in which decomposition results in no harmful by-products); and doing chemistry in water (using water as solvent).

Biomimicry 3.8

The Biomimicry Institute is a non-profit organisational proponent of biomimicry at the forefront of propagating this novel field as well as the brain behind Askna-ture and Biomimicry 3.8. As stated on the website, Biomimicry 3.8 is acclaimed to be the world's leading nature-inspired consultancy, offering services such as training for professionals and students, inspiration, and biological intelligence con-sulting (Biomimicry 3.8, 2016). It is the business arm of the Biomimicry Institute with major multinational clients and collaborators such as Arup Engineers, Boe-ing, Coca-Cola Company, Colgate-Palmolive, Environmental Protection Agency (EPA), General Electric, Hewlett-Packard, HOK Architects, Interface, Johnson & Johnson, Kohler, NASA, Nike, Procter & Gamble, and Shell, amongst others.

Janine Benyus, one of the co-founders of Biomimicry 3.8, clarifies that bio-mimicry is not nature exploitation or extraction but rather, all about learning from nature. This is the reason why it is strongly suggested that multidisciplinary col-laboration whereby biologists and nature conservationists are involved or at the design table, is believed to be key to the successful application of biomimicry. With the developed biomimicry training programmes and methodologies employed across sectors globally, Biomimicry 3.8 can train, equip, and connect sustainability-minded change-makers with nature over 3.8 billion years of innovative and out-standing strategies.

Asknature: Nature-Inspired Database

Several barriers are identified to hinder the adoption and implementation of biomimicry. Five major barriers, namely language barriers (lack of understanding of approaches); integration barriers (lack of biomimicry integration knowledge); environmental policies and principles barriers (non-application of biomimicry principles); conceptualisation barriers (inability to interpret biomimicry principles); and ecosystem complexities barriers (lack of understanding of nature's processes and strategies) are listed by Gamage and Hyde (2012). Vincent et al. (2006) and Zari (2007) identified the lack of a well-defined biomimicry approach as a major barrier to its application. It is for this reason that efforts are made by the Biomimicry Institute to create a special database to document nature-inspired ideas and innovative breakthroughs. This is believed will propagate the concept of biomimicry and also aid the widespread adoption and application.

AskNature is a database of bio-inspired intelligence organised by engineering, design, and scientific functions. As described by Alexandridis (2016), AskNature is the online library established by the Biomimicry Institute to provide a comprehensive and accessible database of nature-inspired design strategies, systems, and functions. Figure 5.4 shows the search interface of AskNature from www.asknature.org. Since its creation, AskNature has recorded over 1700 entries of biological strategies and inspired ideas rooted in nature, making it the world's most accessible and comprehensive online catalogue of nature's solutions to human challenges. With the need to maintain global recognition and address the numerous issues facing humanity, the online library is growing with constant additions being made to the biological strategies, inspired ideas, resources, and collections relative to human innovation challenges. To maximally utilise the online library and gain a high-level understanding, entries are organised by functions which then display

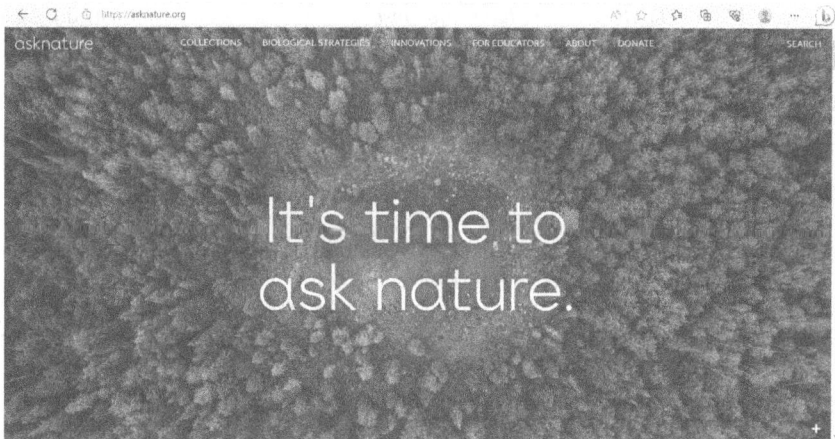

Figure 5.4 Search interface of asknature database

biological strategies and ideas with illustrative media, plain language summaries, and linked references (The Biomimicry Institute, 2018a).

Summary

This chapter provided an overview of the evolution, definitions, and different terminologies that are used interchangeably with biomimicry. The characteristics and dimensions of biomimicry for optimising sustainability are also presented in this chapter. An analysis of the biomimicry application views, design strategies, and approaches for its effective application is also presented in this chapter. Also, the principles of biomimicry that ensure its application results in truly sustainable solutions and innovations are presented in this chapter. This is to serve as a handy checklist for biomimicry applications for sustainability. Finally, an overview of Biomimicry 3.8 and the Asknature database was introduced. The next chapter reviews the fundamentals of sustainability assessment tools in the construction industry.

References

Alexandridis, G. (2016). *Sustainable product design inspired from nature*. Thessaloniki, Greece: International Hellenic University.

Arnarson, P. O. (2011). *Biomimicry: New technology*. Iceland: Reykjavík University.

Ásgeirsdóttir, S. A. (2013). *Biomimicry in Iceland: Present status and future significance*. University of Iceland.

Attenborough, D., & Graham, N. (1995). *The private life of plants: A natural history of plant behaviour*. Princeton, NJ: Princeton University Press.

Aziz, M. S., & El Sheriff, A. Y. (2015). Biomimicry as an approach for bio-inspired structure with the aid of computation. *Alexandria Engineering Journal, 55*(1), 704–714.

Badarnah, L., & Kadri, U. (2015). A methodology for the generation of biomimetic design concepts. *Architectural Science Review, 58*(2), 120–133.

Benyus, J. M. (1997). *Biomimicry: Innovation inspired by nature* (Adobe Digital ed.). Australia: HarperCollins.

Benyus, J. M. (2011). *A biomimicry primer* Biomimicry 3.8.

Bhushan, B. (2009). Biomimetics: Lessons from nature—An overview. *Philosophical Transactions. Series A. Mathematical, Physical, and Engineering Sciences, 367*(1893): 1445–1486.

Biomimicry 3.8. (2016). Biomimicry 3.8—Innovation inspired by nature. Retrieved 21 November 2019 from https://www.biomimicry.net/

Biomimicry, 3. 8. (2017). *SYNAPSE: Spark biomimicry ideas*. Retrieved 11 November 2019 from https://synapse.bio/blog/2017/10/18/free-download-biomimicry-designlens

Browning, W. D., Ryan, C. O., & Clancy, J. O. (2014). *14 patterns of biophilic design: Improving health and well-being in the built environment*. New York NY, USA: Terrapin Bright Green LLC.

Dicks, H., & Blok, V. (2019). Can imitating nature save the planet? *Environmental Values, 28*(5), 519–526.

Eadie, L., & Ghosh, T. K. (2011). Biomimicry in textiles: Past, present and potential. An overview. *Journal of the Royal Society Interface, 8*(59), 761–775.

El Ahmar, S. (2011). Biomimicry as a tool for sustainable architectural design: Towards morphogenetic architecture. (Unpublished master's thesis, Alexandria University).

ElDin, N. N., Abdou, A., & ElGawad, I. A. (2016). Biomimetic potentials for building envelope adaptation in Egypt. *Procedia Environmental Sciences, 34,* 375–386.

El-Zeiny, R. M. A. (2012). Biomimicry as a problem-solving methodology in interior architecture. *Procedia – Social and Behavioral Sciences, 50,* 502–512.

Fayemi, P. E., Wanieck, K., Zollfrank, C., Maranzana, N., & Aoussat, A. (2017). Biomimetics: Process, tools and practice. *Bioinspiration & Biomimetics, 12*(1), 011002.

Ferwati, M. S., AlSuwaidi, M., Shafaghat, A., & Keyvanfar, A. (2019). Employing biomimicry in urban metamorphosis seeking for sustainability: Case studies. *Architecture, City
and Environment (ACE), 14*(40), 133–162.

Fewell, J. H. (2015). Social biomimicry: What do ants and bees tell us about organization in
the natural world? *Journal of Bioeconomics, 17*(3), 207–216.

Freitas, J., & Leitão, A. (2019). *Back to reality.* 24th International Conference of the Association for Computer-Aided Architectural Design Research in Asia (CAADRIA), 173–182.

Gamage, A., & Hyde, R. (2012). A model based on biomimicry to enhance ecologically
sustainable design. *Architectural Science Review, 55*(3), 224–235.

Garcia, P. R. (2017). The influence of the concepts of biophilia and biomimicry in contemporary architecture. *Journal of Civil Engineering and Architecture, 11*(5), 500–513.

Goss, J. (2009). *Biomimicry: Looking to nature for design solutions.* Washington DC: Corcoran College of Art Design.

Graeff, E., Maranzana, N., & Aoussat, A. (2019). Engineers' and biologists' roles during
biomimetic design processes, towards a methodological symbiosis. *Paper Presented at
the Proceedings of the Design Society: International Conference on Engineering Design,
1*(1), 319–328.

Graeff, E., Maranzana, N., & Aoussat, A. (2018). Role of biologists in biomimetic design
processes: Preliminary results. *DS 92: Proceedings of the DESIGN 2018 15th International Design Conference,* 1149–1160.

Gruber, P., & Benti, D. (2013). Biomimetic strategies for innovation and sustainable development. *Sustainable Building Conference SB13,* 1–14.

Hargroves, K., & Smith, M. (2006). Innovation inspired by nature: Biomimicry. *Ecos,
2006*(129), 27–29.

Harkness, J. M. (2004). An idea man (the life of Otto Herbert Schmitt). *IEEE Engineering
in Medicine and Biology Magazine, 23*(6), 20–41.

Helms, M., Vattam, S. S., & Goel, A. K. (2009). Biologically inspired design: Process and
products. *Design Studies, 30*(5), 606–622.

Hoagland, M. B., & Dodson, B. (1999). *Patterns: Sixteen things you should know about life*
(2nd ed.). Toronto, Canada: Jones and Bartlett Publishers.

Kellert, S. (2014). Biophilia and biomimicry: Evolutionary adaptation of human versus nonhuman nature. *Intelligent Buildings International, 8*(2), 1–6.

Kennedy, E., Fecheyr-Lippens, D., Hsiung, B., Niewiarowski, P. H., & Kolodziej, M. (2015).
Biomimicry: A path to sustainable innovation. *Design Issues, 31*(3), 66–73.

Kenny, J., Desha, C., Kumar, A., & Hargroves, C. (2012). Using biomimicry to inform urban
infrastructure design that addresses 21st-century needs. *1st International Conference on
Urban Sustainability and Resilience: Conference Proceedings,* 1–13.

Knippers, J., & Speck, T. (2012). Design and construction principles in nature and architecture. *Bioinspiration & Biomimetics, 7*(1), 1–10.

Lee, S., & Baek, J. S. (2019). Nature-inspired design for self-organized social systems:
A tool for collaborative communities. *Proceedings of the Design Society: International
Conference on Engineering Design, 1*(1), 189–198.

McDonough, W., & Braungart, M. (2010). *Cradle to cradle: Remaking the way we make
things.* London: MacMillan.

Merrill, C. L. (1982). *Biomimicry of the dioxygen active site in the copper proteins hemocyanin and cytochrome oxidase*. Houston, Texas: Rice University.

Murr, L. E. (2015). Biomimetics and biologically inspired materials. *Handbook of materials structures, properties, processing and performance* (pp. 521–552). Cham, Switzerland: Springer.

Neill, P. (2018). *Biomimicry principles: Providing a framework for the future*. Retrieved 11 November 2019 from https://www.huffpost.com/entry/biomimicry-principles-providing-a-framework-for-the_b_5a5534dce4b0e3dd5c3f8ce4

Niewiarowski, P. H., & Paige, D. (2011). *A PhD in biomimicry: How would nature do that?* Proceedings of the First Annual Biomimicry in Higher Education Webinar. the Biomimicry Institute, 14–21.

Nkandu, M. I., & Alibaba, H. Z. (2018). Biomimicry as an alternative approach to sustainability. *Architecture Research, 8*(1), 1–11.

Okuyucu, C. (2015). Biomimicry based on material science: The inspiring art from nature. *Matter, 2*(1), 49–53.

Pandremenos, J., Vasiliadis, E., & Chryssolouris, G. (2012). Design architectures in biology. *Procedia CIRP, 3*, 448–452.

Pólit, J., & David, C. (2014). *Regreening nature: Turning negative externalities into opportunities*. Netherlands: TU Delft, Delft University of Technology.

Pronk, A., Blacha, M., & Bots, A. (2008). Nature's experiences for building technology. *6th International Seminar of the International Association for Shell and Spatial Structures (IASS) Working Group*, 1–11.

Reap, J., Baumeister, D., & Bras, B. (2005). Holism, biomimicry and sustainable engineering. *ASME 2005 International Mechanical Engineering Congress and Exposition*, 423–431.

Reed, P. A. (2004). A paradigm shift: Biomimicry. *The Technology Teacher, 63*(4), 23–28.

Ryan, C. O., Browning, W. D., Clancy, J. O., Andrews, S. L., & Kallianpurkar, N. B. (2014). Biophilic design patterns: Emerging nature-based parameters for health and well-being in the built environment. *International Journal of Architectural Research: ArchNet-IJAR, 8*(2), 62–76.

Scheffer, O. (2016). *From biomimicry to ecomimicry: Reconnecting cities – and ourselves – to earth's balances*. Retrieved 10 November 2019 from https://www.thenatureofcities.com/2016/11/30/from-biomimicry-to-ecomimicry-reconnecting-cities-and-ourselves-to-earths-balances/

Shu, L. H., Ueda, K., Chiu, I., & Cheong, H. (2011). Biologically inspired design. *CIRP Annals-Manufacturing Technology, 60*(2), 673–693.

Singh, A., & Nayyar, N. (2015). Biomimicry – an alternative solution to sustainable buildings. *Journal of Civil and Environmental Technology, 2*(14), 96–101.

Speck, T., & Speck, O. (2008). Process sequences in biomimetic research. *WIT Transactions on Ecology and the Environment, 114*, 3–11.

Stafford, J. (2011). *Steve Jobs and the next big "intersection"*. Retrieved 10 November 2019, from https://scopeblog.stanford.edu/2011/11/02/steve-jobs-and-the-next-big-intersection/

Taylor Buck, N. (2015). The art of imitating life: The potential contribution of biomimicry in shaping the future of our cities. *Environment and Planning B: Planning and Design, 44*(1), 120–140.

Tempelman, E., Pauw, I. C. D., Grinten, B. V. D., Mul, E. J., & Grevers, K. (2015). Biomimicry and cradle to cradle in product design: An analysis of current design practice. *Journal of Design Research, 13*(4), 326–344.

The Biomimicry Institute. (2013a). *Biomimicry Design Lens*. Retrieved 03 November 2019 from https://asknature.org/resource/biomimicry-designlens/#.XnXSkqgzbIU

The Biomimicry Institute. (2013b). *The biomimicry taxonomy: Biology organized by function*. Retrieved 20 March 2020 from http://toolbox.biomimicry.org/wp-content/uploads/2015/01/AN_Biomimicry_Taxonomy.pdf

The Biomimicry Institute. (2018). *AskNature*. Retrieved 21 November 2019 from https://asknature.org/

Vattam, S., Helms, M. E., & Goel, A. K. (2007). *Biologically-inspired innovation in engineering design: A cognitive study*. USA: Georgia Institute of Technology.

Vincent, J. F., Bogatyreva, O. A., Bogatyrev, N. R., Bowyer, A., & Pahl, A. (2006). Biomimetics: Its practice and theory. *Journal of the Royal Society Interface, 3*(9), 471–482.

Volstad, N. L., & Boks, C. (2008). Biomimicry – A useful tool for the industrial designer? *DS 50: Proceedings of Nord Design 2008 Conference*, Tallinn, Estonia, 21.23.08. 2008.

Volstad, N. L., & Boks, C. (2012). On the use of biomimicry as a useful tool for the industrial designer. *Sustainable Development, 20*(3), 189–199.

Zari, M. P. (2007). Biomimetic approaches to architectural design for increased sustainability. *Lisbon Sustainable Building Conference (SB07)*, 1–10.

Zari, M. P., & Storey, J. B. (2007). An ecosystem based biomimetic theory for a regenerative built environment. *Sustainable Building Conference*, 7.

Zhang, M., Gu, Z., Bosch, M., Perry, Z., & Zhou, H. (2015). Biomimicry in metal – organic materials. *Coordination Chemistry Reviews, 293*, 327–356.

Part III

Sustainability Assessment and Rating in Construction

6   Sustainability Assessment Tools in the Construction Industry

Introduction

The emergence of sustainability assessment is a direct effect of the concept of sustainability (Pope et al., 2004), especially in the construction industry (CI). Owing to the adverse environmental impacts attributed to construction activities globally, it has become an important tool in the pursuit of sustainability in the CI. According to Berardi (2012), the use of sustainability assessment tools (SATs)/systems has grown globally with the only exceptions being Latin America (except Brazil) and Africa (except South Africa). In literature, different terms have been used to refer to sustainability assessment as it pertains to the industry. Despite being used interchangeably, the terms Sustainability Assessment Methods/SATs (Kumanayake & Luo, 2017), Building Environmental Assessment (Gu et al., 2006), Building Sustainability Assessment Methods (Markelj et al., 2014), Environmental Assessment Tools (Haapio & Viitaniemi, 2008), Sustainability Certification Schemes (Siebers et al., 2016), Building Environmental Assessment Tools/Scheme (Sev, 2011; Weerasinghe, 2012; Lee, 2013; Ng et al., 2013; Wallhagen et al., 2013), Environmental Building Assessment Tools/Methods (Ding, 2008; Ali & Al Nsairat, 2009), Neighbourhood SATs (Yıldız et al., 2016), Sustainable Building Rating Systems (Fowler & Rauch, 2006), Green Building Rating System (CSI, 2013; Maheepala et al., 2017), Green Building Certification System (Said & Harputlugil, 2019), Rating Systems, Standards and Guidelines (RSMeans, 2011), Sustainability/Environmental Rating Systems (Poveda & Lipsett, 2011; Jayawickrama, 2014), Building Sustainability Assessment (Carvalho et al., 2019), Building Assessment System (Presley & Meade, 2010), Environmental Labels (Fuerst & McAllister, 2011) and Sustainability Evaluation Systems (Yu et al., 2018) all describe the evaluation of a building or project to ascertain its compliance with sustainable practices.

Overview of Sustainability Assessment Tools

SATs/methods in the CI are developed with the sole aim of assessing the sustainability performance of engineering and building construction projects against defined sustainability criteria, resulting in a label for such buildings (Huang & Hsu, 2011; Siebers et al., 2016). However, this chapter interchangeably adopts the terms

DOI: 10.1201/9781003415961-9

SATs/methods and green building rating tools (GBRTs) for consistency. SATs/methods in the CI are divided into three categories according to their specific functions, namely the assessment method, rating system and labelling system which are differentiated as follows:

- The assessment method comprises a set of criteria (generally life cycle assessment (LCA)-based) used in assessing the sustainability performance of a building/project;
- The rating system is an evaluation procedure for each performance issue area, with the output in scores and;
- A labelling system is a marketing programme to publicise a project to the industry and is implemented by expert assessors (Larsson, 2004).

According to Ding (2008), assessment (evaluation) tools/methods proffer quantitative performance benchmarks for design alternatives while rating tools/systems establish the performance level of a building in stars. The application of sustainability indicators can be for signalling/marketing, labelling, design for environment/design for sustainability, sustainability reporting, social responsibility, green procurement/sustainability procurement, and periodic review (ISO, 2011). However, most of the existing tools combine the functions of assessment and rating to provide comprehensive information on the building project.

Although sustainability is a complex task to achieve, it has been found that construction stakeholders have concentrated mainly on environmental objectives with much focus on human health and the environment (Hes, 2007). The study of Said et al. (2009) attested to the fact that most of the existing sustainability assessment methods only consider the environmental elements, leaving out the other parts. Relatively few of the SATs/methods integrate the principles of the triple bottom line of sustainability (Goh & Rowlinson, 2013). While most of these assessment tools/systems focus on addressing and evaluating the environmental performance indicators, few focus on the social and economic aspects instead of addressing the combination of environmental, social, and economic performance indicators. A remarkable example is the international sustainable building tool (SBTool) and South Africa's sustainable building assessment tool (SBAT) which are believed to be comprehensive as they encompass a synergy of the three sustainability dimensions. However, most of the existing ones predominantly focus on the environmental aspect of sustainability, thereby leaving out the social and economic dimensions, and hence failing to pursue the goals of sustainability holistically. Despite this disparity in focus, all the existing assessment tools/methods share a common goal of stimulating market demands for the increased environmental performance of buildings/projects (da Silva, 2007).

Sustainability Assessment Tools: Definitions

Basically, a building/project can be adjudged as green or as performing sustainably without the application of any SATs/system to it. However, SATs offer a mechanism for measuring, supplying, recognising, and validating a project's level of adherence to sustainable processes (RSMeans, 2011). Definition of Siebers et al. (2016) described SATs/systems support systems for sustainable planning, constructing and operating building projects as well as a quality assurance tool. Also described as benchmarking tools, SATs assess a project along a set of criteria, rate its performance against a given set of criteria values/ standards, and lastly, communicate a value judgement about its performance (Herda et al., 2017). They are effective tools available to stakeholders in the industry for policy and decision-making processes (Saad et al., 2019). According to Waas et al. (2014), sustainability assessment is a process that aims to achieve the following:

- Contribute to the better comprehension of the meaning and contextual interpretation of sustainability;
- Incorporate sustainability issues and principles into decision-making by identifying and evaluating (past and/or future) sustainability impacts; and
- Promote the objectives of sustainability.

Further evidence showed that they are tools that appraise the performance or expected performance of a building/project and further interpret such an appraisal into a comprehensive evaluation for comparison against other buildings/projects (Fowler & Rauch, 2006). These rating systems offer standards and guidelines for measuring how green a building is while also supplying recognition and validation of such claims (RSMeans, 2011). They are schemes that provide a sturdy framework used in assessing the sustainability performance of buildings/projects (proposed and existing) and promoting the design, implementation, adoption, and proliferation of GBs (Sebake, 2008; Da Silva, 2015). These tools help in addressing environmental issues during the construction processes and assessing the environmental performance of buildings. They also help improve the construction market and sphere in terms of meeting the goals of green projects. They are voluntary tools developed in each country/region to support the development of GBs, apart from the Sustainable Building Challenge and Living Building Challenges which are international rating standards. However, most of these assessment tools are country-specific, while some are adopted and applied outside their country of origin.

By satisfying specified sustainability criteria, the use of sustainability assessment systems/methods aids the project certification process (Goh & Rowlinson, 2013). These tools help in understanding the eco-friendliness of a building/project and assist in identifying areas for improvement to minimise the environmental impacts (Ng et al., 2013). In summary, a SATs/system should evaluate cogent technical aspects of sustainability in a project lifecycle, offer valid and unvarying premises for comparison, and must not show any form of complexity in its communication and implementation (Fowler & Rauch, 2006). It is then that such a tool will deliver on the wider agenda of sustainability and beyond the popular concentration on environmental responsiveness as observed in most currently used tools.

Taxonomy of Sustainability Assessment Tools

In the literature, there are different ways by which SATs have been classified. According to the study of Mahmoud et al. (2019), some of these assessment tools/systems are categorised based on their performance (i.e., multi-attribute assessment or life-cycle assessment tools), while others are based on their scope (multi-aspects or single aspect of the building project). From the typology of SATs presented by Gasparatos and Scolobig (2012), they can be classified as monetary tools (valuation and aggregation tools), biophysical tools (ecological footprint, energy synthesis, material flow auditing, exergy analysis), and indicator tools (multi-criteria assessment, composite indicators). The study of Ali and Al Nsairat (2009) affirmed that SATs can either be (a) criteria-based, or (b) LCA methodology-based. The criteria-based SATs involve assigning points to determined criteria, examples of which include BREEAM and LEED, amongst others. The LCA methodology-based SATs (i.e., EcoQuantum, KCL Eco) on the other hand evaluate the impact of a building/project (including the process and products) during its lifetime (cradle to grave).

According to the International Organisation for Standardisation (ISO), sustainability assessment indicators are classified according to their functions, namely communication, simplification, and quantification (ISO, 2011). A three-level classification of SATs called the Assessment Tool Typology was developed by the ATHENA Sustainable Materials Institute. The study by Trusty (2000), and Haapio and Viitaniemi (2008), offer a basic framework for assessment tool comparison and are as follows:

- Level 1: Product comparison tools and information sources (i.e., BEES 3.0, TEAMTM).
- Level 2: Whole building decision support tools (EcoQuantum, ATHENA, Envest).
- Level 3: Whole building assessment frameworks or systems (GBTool, BREEAM, LEED).

Level 1 tools are beneficial for building databases and making comparisons and choices at the procurement stage. Level 2 tools involve weighing or scoring and focusing on specific (or combined) areas of concern such as operating energy, life-cycle cost, and lighting among others. Level 3 tools provide a broad coverage of the three sustainability pillars, using a mix of subjective and objective data to distil the information and provide usable overall measures.

Along similar lines, Jawali and Fernández-Solís (2008) suggested that SATs can also be categorised into two according to their usage in the CI, namely assessment and rating tools. While the assessment tools aim to achieve sustainability by providing quantitative performance indicators, rating tools are aimed at determining the performance level of a building project in graphics (Stars). Despite being established and operated by either government or private voluntary organisations, SATs aim to achieve the core objective of sustainability in building projects by providing a comprehensive set of requirements or areas of compliance to be adhered to.

Annex 31, a project under the auspices of the International Energy Agency (IEA), also provided a classification of SATs into five categories. The classification integrated the three levels developed by the ATHENA Sustainable Materials Institute. As highlighted by Haapio and Viitaniemi (2008) and Goodhew (2016), they are as follows:

- Class 1: Energy modelling software
- Class 2: Environmental LCA tools for buildings and building stocks (comprises of levels 1, 2, and 3 of the ATHENA classification. i.e., BEES 3.0, EcoQuantum, EcoEffect)
- Class 3: Environmental assessment frameworks and rating systems (Level 3 of the ATHENA classification. i.e., BREEAM and LEED)
- Class 4: Environmental guidelines or checklists for design and management of buildings
- Class 5: Environmental product declarations, catalogues, reference information, certifications, and labels.

SATs/systems can also be grouped as LCA, total quality assessment (TQA), cumulative energy demand, and methodology of architecture towards sustainability (ATS) (Hastings & Wall, 2007; Berardi, 2012; Mahmoud et al., 2019). The CED category focuses on measuring and evaluating the building's energy consumption, the LCA category measures and evaluates the environmental aspects, the TQA category aims at measuring and evaluating the social, economic, and ecological aspects, while ATS evaluates the social, economic, environmental, and political dimension of a building. RSMeans (2011) classified SATs into two groups, namely those that relate to the building/project as an unabridged entity (whole building multiple attribute ratings) and those that relate to specific building components (whole building single attribute ratings). The whole building multiple attribute rating system considers and evaluates the sustainability performance of a building/ project in its entirety. Examples of such tools are LEED, CASBEE, BREEAM, and GBTools, among many others. However, the whole building's single attribute rating system is limited to the assessment of specific properties of buildings, materials, and systems. Examples of such tools are Energy Star and WaterSense, amongst others.

Another classification put forward in a report by the Organisation for Economic Co-operation and Development (OECD, 2003) stated that SATs can be mandatory energy labelling, voluntary environmental labelling, and voluntary comprehensive labelling. Mandatory energy labelling, an obligatory scheme, evaluates the energy efficiency of a building and furnishes potential buyers or users with the result. Voluntary environmental labelling, known as the most common and voluntary type of labelling scheme, evaluates a building against a wide range of environmental issues such as energy, material use and indoor air quality, among others. BREEAM is a typical example of an assessment scheme under this category. Lastly, voluntary comprehensive labelling is a voluntary assessment scheme that evaluates buildings against environmental criteria and other basic issues. An example of scheme under this category is the Haute Qualité Environnementale

Scheme in France and the Housing Performance Indication Scheme introduced in 2001 in Japan.

In a study where the most widely adopted rating systems for evaluating the environmental impacts of buildings are analysed, Bernardi et al. (2017) classified SATs under four headings, namely Categories (a specific set of environmental performance issues considered during assessment); Scoring system (a performance evaluation system that sums up the number of credit points earned when a given level of conformity to analysed aspects is achieved); Weighing system (the relevance attached to a specific category within the overall scoring system); and Output (shows a comprehensive and direct result of the achieved environmental performance during the scoring stage).

However, the main functions of SATs as presented by Gu et al. (2006) can be summarised as follows:

- Promote the uptake of regulations and policies focussing on sustainable construction (SC);
- Present a set of quantitative environmental measures and evaluating methods for achieving sustainable buildings;
- Improve the environmental management of the built environment;
- Task the construction industry to pay rapt attention to environmental issues;
- Raise awareness of environmental standards and issues; and
- Recognise and encourage environmental design considerations for buildings while stimulating the green market.

Core Attributes of Sustainability Assessment Tools

Most of the existing SATs are country-specific; however, they all still play a cogent part in raising awareness and popularising the green movement in the CI. However, for an assessment tool to perform its function, RSMeans (2011) advocated these must be science-based (results and decisions must be reproducible by others using the same tool); transparent (process and standards leading to the certification should be open for examination and transparent); objective (conflict-free certification body); and progressive (the tool should advance industry practices). The study of Jayawickrama (2014) identified comprehensiveness (assessment of a broad range of environmental issues and criteria); weighing systems (prioritising environmental issues and criteria by allocating different numbers of assessment points to different criteria); results communication (presenting performance benchmarks in a coherent and informative way); and qualitative criteria (comprehensive evaluation using qualitative metrics).

Each of the assessment tools lay out a set of absolute criteria to be adhered to in reaching a threshold target of sustainability compliance. To be able to evaluate how sustainable a building/project is, it is important for an SAT to holistically consider and integrate the following principles:

- Building envelope design that minimises adverse environmental impact;
- Water efficiency and waste management;

- Sustainable site planning;
- Integration of renewable energy sources such as wind, solar, and alternative energy to generate energy onsite;
- Building system design that integrates high-performance and energy-efficient HVAC (heating, ventilation, and air conditioning), lighting, electrical, and water heating systems;
- Indoor environmental quality to maintain indoor air quality and thermal and visual comfort; and
- Use of ecologically sustainable materials and products that possess high recycled content, are rapidly renewable and have minimum off-gassing of harmful chemicals (Kubba, 2012).

Ching and Shapiro (2014) listed site selection and sustainability, water efficiency and conservation, material and resource selection, energy efficiency and conservation, indoor environmental quality, historical and cultural significance, acoustics, safety and security, and beauty as cogent criteria assessment tools should consider. Reeder (2010) also identified site selection, site development, energy efficiency, water conservation, material and resource efficiency, indoor environmental quality, owner/tenant education, innovation and design, and global impact. According to Kaur and Garg (2019), a typical SATs/system should include waste management, innovation (technology, design), transportation, site (location, linkage, planning, sustainability), land use planning, resources (energy, water, material), social and economic well-being, and infrastructure as the major categories according to which a development should be evaluated. Described as SC index/indicators, the study of Huang and Hsu (2011) established a five-layer system that consists of the indicator, indicator category, core cluster, theme, and overall performance. Table 6.1 presents the constituents of the SC index/indicators.

Akadiri and Olomolaiye (2012) categorised the sustainable performance criteria as environmental (minimise pollution, zero/low toxicity, ozone depletion potential, environmental statutory compliance, recyclable/reusable material, amount of likely wastage in use, embodied energy in material, impact on air quality, impact during harvest, methods of extraction of raw materials, environmental sound disposal options); technical (maintainability, fire resistance, resistance to decay, energy saving and thermal insulation, ease of construction/buildability, and life expectancy/durability); and socio-economic criteria (aesthetics, health and safety, first cost, disposal cost, use of local materials, labour availability, and maintenance cost). The study of Whang and Kim (2015) categorised the key factors according to the triple bottom line of sustainability, namely economic issues (life cycle cost, quality management for durability, knowledge management, retention of skilled labour, project delivery, partnering, value for money in the delivery, competitiveness, productivity/profitability, construction cost, commercial viability, operating and maintenance cost, affordability of cost levels, image and reputation, support of local economy, innovation/research and development); social issues (service quality, security, community, well-being, employment, health and safety, partnership working, culture/heritage, education/training); and environmental issues (management, transport, ecological environment,

Table 6.1 Sustainable construction index

Overall performance	Theme	Core cluster	Indicator category	Indicator
Sustainable Construction Index	Environment	Resource Usage	Land Materials Water Energy	Individual indicators are local constraints, specific features, and national priorities
		Pollution reduction	Air Water Solid waste Noise Toxic substance GHG emissions	
		Ecological protection	Biodiversity Ecologically sensitive areas Replanting of vegetation and afforestation in construction	
	Society	Improving the life of mankind	Quality of human settlement Transportation Health, safety, security, and threats	
		Conservation of culture	Preservation of cultural heritage Accepting cultural differences	
		Social equity	Equality of regional development Equitable distribution of resources, social costs, and benefits of construction Gender equality Universal design	
		Social aspects of the construction process	Stakeholder involvement Quality of construction Avoidance of child labour CSR of construction companies	
	Economy	Economic contribution	Contribution towards the creation of employment Contribution towards the growth of national economy	
		Eco-economics	Production of green building materials Subsidies	
		A measure of economic efficiency	Benefits/costs of construction project Value added to land and buildings Construction company profits	

Source: Huang and Hsu (2011).

atmosphere, indoor environmental quality, use of land, water efficiency, energy efficiency, and material and resources).

The study of He et al. (2019) on product sustainability assessment identified five indicators for comprehensive evaluation. These include resource (material resource, material utilisation rate, equipment efficiency, equipment usage, equipment resource, equipment failure rate); energy (energy efficiency, energy usage, energy consumption, clean energy usage rate); economic (cost, recyclable rate); environmental (solid waste generation, wastewater discharge, exhaust gas emissions); and technical indicators (reliability, precision, product configuration). Understanding that sustainability is a holistic and dynamic process, the study of Latif et al. (2017) in developing a sustainability index for the manufacturing industry listed waste management, energy efficiency, workers' health, and safety as important indicators to be evaluated. A conceptual framework for implementing sustainability in building construction by Akadiri et al. (2012) highlighted the following based on the three dimensions of SD, namely energy conservation, material conservation, water conservation, and land conservation (Resource Conservation); initial cost (purchase cost), cost in use, and recovery cost (Cost Efficiency); and protecting physical resources and protecting human health and comfort (Design for Human Adaptation). However, despite the numerous classifications of SATs/systems by various researchers, Bhakar et al. (2018) suggested a well-balanced incorporation of all the indicator tools to achieve the overarching goal of sustainability.

Forty-three categories which are categorised under the issues of sustainable site, water efficiency, energy efficiency, material and resources, indoor environment quality, environment loading, social and economic aspects, cultural aspects, and service quality are identified as crucial for sustainable building assessment by the study of Bhatt and Macwan (2012). In tandem with the United Nation's 5Ps' (progress, people, planet, prosperity, and place) categorisation of SD attributes, the study by Gusta (2016) highlighted that the determination of the sustainability performance of a building project should be evaluated based on the following:

a   Progress dimension: innovation and transferability;
b   People dimension: ethical standards and social inclusion;
c   Planet dimension: resource and environmental performance:
d   Prosperity dimension: economic viability and compatibility; and
e   Place dimension: contextual and aesthetic impact.

International Standards for Sustainability in Buildings and Civil Engineering Works

The Global Federation of National Standard Bodies known as the ISO is saddled with the task of developing, preparing, and publishing international standards. This assignment is carried out by a technical committee which consists of each member body interested in the subject for which a standard is required. Governmental and non-governmental bodies, international organisations and other relevant stakeholders in liaison with the technical committee of the ISO collaborate

Table 6.2 Numbers of ISO standards applicable to SDGs

Goals	Description	Number of ISO standards applicable
Goal 1	No Poverty	78
Goal 2	Zero Hunger	70
Goal 3	Good Health and Well-being	393
Goal 4	Quality Education	87
Goal 5	Gender Quality	54
Goal 6	Clean Water and Sanitation	171
Goal 7	Affordable and Clean Energy	162
Goal 8	Decent Work and Economic Growth	184
Goal 9	Industry, Innovation and Infrastructure	570
Goal 10	Reduced Inequalities	131
Goal 11	Sustainable Cities and Communities	286
Goal 12	Responsible Consumption and Production	314
Goal 13	Climate Action	231
Goal 14	Life Below Water	103
Goal 15	Life on Land	190
Goal 16	Peace, Justice, and Strong Institutions	68
Goal 17	Partnerships for the Goals	0

Source: ISO (2020).

closely on the task of developing and publishing international standards with over 22,000 standards and documents published to date. Of the numerous standards prepared by the ISO, there are a few that address sustainability in the building CI. Table 6.2 also describes the number of ISO standards that directly contribute to the SDGs.

ISO 21929-1 under the general title 'Sustainability in Building Construction – Sustainability Indicators' consists of part 1 which is the framework for the development of indicators and a core set of indicators for buildings, while part 2 is still under development. According to ISO (2011), ISO 21929-1, Sustainability in building construction – sustainability indicators is one out of the five international standards that address sustainability in the CI. The other four include ISO 15392 (Sustainability in building construction – General principles); ISO/TR 21932 (Building construction – Sustainability in building construction – Terminology); ISO 21930 (Sustainability in building construction – Environmental declaration of building products); and ISO 21931-1 (Sustainability in building construction – Framework for methods of assessing the environmental performance of construction works – Part 1: Buildings). As shown in Table 6.3, ISO 21929-1, ISO 15392, and ISO/TR 21932 encompass the environmental, economic, and social aspects of sustainability while ISO 21930 and ISO 21930 encompass only the environmental aspect. It is therefore imperative to note a major trend and similarity in ISO 15392, ISO 21929-1, ISO 21930, ISO 21931-1, ISO 21932, and other SATs which is their inability to address all 17 SDGs thereby authenticating the need to create an assessment tool that addresses these shortcomings and others as well.

Table 6.3 Relationship among international standards for sustainability in buildings and civil engineering works

	Social dimension	*Economic dimension*	*Environmental dimension*
Methodological basics	**ISO 15392**: Sustainability in building construction – General principles **ISO/TR 21932**: Building construction – Sustainability in building construction – Terminology **ISO 21929-1**: Sustainability indicators – Part 1: Framework for the development of indicators and a core set of indicators for buildings		
Buildings			**ISO 21931-1**: Sustainability in building construction – Framework for methods of assessment environmental performance of construction works – Part 1: Buildings
Building products			**ISO 21930**: Sustainability in building construction – Environmental declaration of building products

Source: ISO (2008).

ISO 21929-1: Sustainability in Building Construction – Sustainability Indicators

ISO 21929-1 was published in 2011 and a review of the standard was confirmed in 2017. According to ISO (2011), this part of ISO 21929 is about a 'framework for the development of sustainability indicators for buildings based on the premise that sustainable development of buildings brings about the required performance and functionality with minimum adverse environmental impact while encouraging improvements in economic and social (and cultural) aspects at local, regional and global views'. This standard provides the criteria for developing sustainability standards for buildings and clarifies the building features for consideration. As stated on the official website, ISO 21929-1 contributes to only nine sustainable development goals (SDGs), namely 3, 6, 7, 8, 10, 11, 13, 14, and 15 (ISO, 2020).

As against the generally known triple pillars of SD which are social, economic, and environmental, the core areas of protection for ISO 21929-1 are seven, namely ecosystem, natural resources, health and well-being, social equity, cultural heritage, economic prosperity, and economic capital. As listed by ISO (2011), the following 14 major aspects of the building impact the core areas of protection with their corresponding representative indicators:

a   Emissions to air: global warming potential (GWP) and ozone depletion potential;
b   Use of renewable resources (amount of non-renewable resources consumption by type): consumption of non-renewable raw materials and consumption of non-renewable energy;
c   Freshwater consumption (amount of freshwater consumption);
d   Waste generation (amount of waste generation by type);

e   Change of land use;

f   Access to services (access to services by type): public modes of transportation, personal modes of transportation, green and open areas, and user-relevant basic services;

g   Accessibility: accessibility of the building site (curtilage) and accessibility of the building;

h   Indoor conditions and air quality: indoor thermal conditions, indoor visual conditions, indoor acoustic conditions, and indoor air quality;

i   Adaptability: change of use or user needs and adaptability for climate change;

j   Costs (life cycle costs);

k   Maintainability;

l   Safety: structural stability, fire safety, and safety in use;

m   Serviceability; and

n   Aesthetic quality.

Sustainability Versus Green Building Rating Systems

The global attention to the need for greening the processes and activities of the CI has resulted in the evolution of numerous SATs/systems. Though imperfect in some ways, each of these tools reflects a deliberate and invaluable commitment towards protecting the environment and human health (Ching & Shapiro, 2014). As observed by Li et al. (2017), three levels of hierarchy, in the following order, namely categories, criteria, and indicators are found to be common to all SATs. The categories' level tops the list of the hierarchy and helps in defining the scope of an SAT. The criteria, which are subsets of the broader categories, are second on the hierarchy table and comprise indicators which aid the proper evaluation of a building's sustainability. This section discusses the SATs that are globally applied in the CI and that fall into the criteria-based, comprehensive, and qualitative categories which is also the target of this study. It is reported that there are more than 30 SATs in use all over the world (Uğur & Leblebici, 2018). Most of these tools are country-specific and utilised by member countries of the World Green Building Council (WGBC). Table 6.4 reveals the list of notable SATs/methods used in the CI, their corresponding country of use, the developer, and their corresponding sources.

Summary

To fully comprehend the fundamentals and concept of assessment tools for buildings, this chapter examined the theoretical foundation of SATs. The chapter presented the overview, definitions, classification, and features of what constitutes a SAT. The international standard for sustainability in buildings and civil engineering works is also discussed to grasp the concept of holistic sustainability in the built environment, which is what this book seeks to achieve. Hence, ISO 21929 was examined as it provides the template for developing sustainability standards for buildings based on the tenets of sustainable development. The next chapter identifies, presents, and examines the various tools, systems, and methods for evaluating GBs globally.

Table 6.4 Sustainability and green building assessment tools

S/N	Sustainability assessment tools/methods	Country of use	Developer	Authors
1	Building Research Establishment Environmental Assessment Method (BREEAM)	UK	Building Research Establishment (BRE), United Kingdom	Lee and Burnett (2008); Sev (2011); Kim et al. (2013); Chen et al. (2015); Doan et al. (2017); Herda et al. (2017); Sharma (2018)
2	Leadership in Energy in Environmental Design (LEED)	USA	United States Green Building Council (USGBC)	Sev (2011); Kim et al. (2013); Chen et al. (2015); Doan et al. (2017); Herda et al. (2017); Sharma (2018)
3	Comprehensive Assessment System for Building Environment Efficiency (CASBEE)	Japan	Japan GreenBuild Council (JaGBC), Institute for Building Environment and Energy Conservation (IBEC) and Japan Sustainable Building Consortium (JSBC)	Sev (2011); Kim et al. (2013); Zhang et al. (2017); Sharma (2018)
4	Leadership in Energy in Environmental Design (LEED Canada Rating System)	Canada	Canada Green Building Council (CaGBC)	CGBC (2020)
5	Deutsche Gesellschaft für Nachhaltiges Bauen (DGNB)	Germany	Germany Sustainable Building Council	Zuo and Zhao (2014); Herda et al. (2017); Zhang et al. (2017)
6	Living Building Certification (LBC)	International	International Living Future Institute (ILFI)	ILFI (2020)
7	GreenStar Australia	Australia	Green Building Council of Australia (GBCA)	Sev (2011)
8	National Australian Built Environment Rating System (NABERS)	Australia	Government of New South Wales (NSW) Department of Planning, Industry and Environment	Huang et al. (2014)
9	Beam Plus (formerly Hong Kong's Building Environmental Assessment Method HK-BEAM)	Hong Kong	Hong Kong Green Building Council and BEAM Society Limited	Lee and Burnett (2008); BEAM Society Limited (2012); Cheng & Venkataraman, 2012); Hui (2017); HKGBC (2020)

(Continued)

Table 6.4 (Continued)

S/N	Sustainability assessment tools/methods	Country of use	Developer	Authors
10	Comprehensive Environmental Performance Assessment Scheme for Buildings (CEPAS)	Hong Kong	Buildings Department, Hong Kong Special Administrative Region (HKSAR) Government	Wu and Yau (2005)
11	Green Star New Zealand	New Zealand	New Zealand Green Building Council (NZGBC)	NZGBC (2019)
12	Green Star South Africa	South Africa	Green Building Council South Africa (GBCSA)	GBCSA (2017)
13	Sustainable Building Assessment Tools (SBAT)	South Africa	Council for Scientific and Industrial Research (CSIR)	Gibberd (2002); Gibberd (2003)
14	Evaluation Standard for Green Building (ESGB)	China	Ministry of Construction	Ma et al. (2016); Zhang et al. (2017); Ding et al. (2018)
15	Assessment Scheme for the Environmental Performance of Buildings (EPB) by Haute Qualité Environnementale (HQE)	France	Haute Qualité Environnementale (HQE) and Cerway (Alliance HQE-GBC)	Cerway (2016)
16	Green Mark Scheme	Singapore	Building and Construction Authority (BCA), Ministry of National Development	Nguyen and Gray (2016); Seghier et al. (2017)
17	ARZ Building Rating System	Lebanon	Lebanon Green Building Council (LGBC)	LGBC (2020)
18	Building Environmental Performance Assessment Criteria (BEPAC)	Canada, British Columbia	University of British Columbia	Kim et al. (2013)
19	Building Sustainability Index (BASIX)	Australia, New South Wales	New South Wales Government	IEA (2020)
20	Çevre Dostu Yeşil Binalar Derneği, House Certificate (ÇEDBİK-Konut Sertifikası)	Turkey	Green Building Association	ÇEDBİK (2017)
21	Code for Sustainable Homes (CFSH)	UK	Department for Communities and Local Government, London, and BREEAM Centre at the Building Research Establishment (BRE)	Crown (2010); UK Building Compliance (2019)
22	LOTUS	Vietnam	Vietnam Green Building Council	Nguyen and Gray (2016)

(*Continued*)

Table 6.4 (Continued)

S/N	Sustainability assessment tools/methods	Country of use	Developer	Authors
23	Building for Ecologically Responsive Design Excellence	Philippines	Philippines Green Building Council	PHILGBC (2015); Nguyen and Gray (2016)
24	Green Building Index	Malaysia	Greenbuildingindex Sdn Bhd, PAM Architects, and Association of Consulting Engineers Malaysia (ACEM)	Nguyen and Gray (2016); GBI (2020)
25	GREENSHIP	Indonesia	Green Building Council Indonesia	Nguyen and Gray (2016); GBCI (2020)
26	Green Standard for Energy and Environmental Design (G-SEED) formerly Korea's Green Building Certification Criteria (KGBCC)	South Korea	South Korea's Ministry of Land, Infrastructure and Transportation (MLIT) and Ministry of Environment	Kim et al. (2013); ESCI (2012); Jeong et al. (2016)
27	Israeli Green Building Standard SI 5281	Israel	Standards Institution of Israel (SII)	Cohen et al. (2017)
28	Green Rating for Integrated Habitat Assessment (GRIHA)	India	The Energy and Resource Institute (TERI), Ministry of New and Renewable Energy, and Leadership in Energy and Environment Design (LEED), operated by the Indian Green Building Council (IGBC)	Sharma (2018)
29	GREEN^SL Rating System	Sri Lanka	Green Building Council of Sri Lanka (GBCSL)	Guanwardana et al. (2017)
30	Pearl Rating System for Estidama	Abu Dhabi	Abu Dhabi Urban Planning Council	Zuo and Zhao (2014)
31	BREEAM-LV	Latvia	Latvian Sustainable Building Council (LSBC)	Gusta (2011); Mishra and Kauškale (2018)
32	BREEAM-NL	Netherlands	Dutch Green Building Council (DGBC)	DGBC (2020); Dekkers (2017)
33	BREEAM-SE	Sweden	Sweden Green Building Council (SGBC)	Freitas and Zhang (2018); Turk et al. (2018); SGBC (2020)

(*Continued*)

Table 6.4 (Continued)

S/N	Sustainability assessment tools/methods	Country of use	Developer	Authors
34	Excellence in Design for Greater Efficiencies (EDGE) Green Building Certification System	Global	International Finance Corporation (IFC) and Green Business Certification Institute (GBCI)	Nguyen et al. (2017); EDGE (2020); GBCI (2020)
35	Green Building Council (GBC) Brasil CASA Certification	Brazil	Green Building Council Brasil (GBCB)	GBCB (2020)
36	CASA Colombia	Colombia	Colombian Council for Sustainable Construction (CCCS)	CCCS (2016)
37	Indian Green Building Council (IGBC) Rating Systems	India	Indian Green Building Council (IGBC)	Adegbile (2013); IGBC (2015); Sampat et al. (2015); Srivastava et al. (2017)
38	WELL Building Standard	International	The International WELL Building Institute (IWBI)	IWBI (2020)
39	Pakistan Green Building Guidelines (PGBG)	Pakistan	Pakistan Green Building Council (PGBC)	PGBC (2019)
40	Home Performance Index (HPI)	Ireland	Irish Green Building Council (IGBC)	IGBC (2020)
41	Miljöbyggnad (Swedish Environmental Certification for Swedish Conditions)	Sweden	Sweden Green Building Council	SGBC (2020)
42	Building Sustainability Assessment Method (BSAM) for Developing Countries in Sub-Saharan Africa	Sub-Saharan Africa (Nigeria)	Timothy Olawumi, Daniel Chan, Albert Chan, and Johnny Wong	Olawumi et al. (2020)
43	Simplified Method for Evaluating Building Sustainability (SMEBS) in the Early Planning Phases for Architects	Global (Slovenia)	Jernej Markelj, Kitek Kuzman Manja, Grošelj Petra, Zbašnik-Senegačnik Martina	Markelj et al. (2014)
44	Building Information Modeling (BIM) based Kazakhstan Building Sustainability Assessment Framework (KBSAF)	Kazakhstan	Gulzhanat Akhanova, Abid Nadeem, Jong R. Kim, Salman Azhar	Akhanova et al. (2020)
45	Green Globes Certification	US, Canada, Global	Green Building Initiative	GBI (2020)

(Continued)

Table 6.4 (Continued)

S/N	Sustainability assessment tools/methods	Country of use	Developer	Authors
46	SBTool	Canada, Global	International Initiative for a Sustainable Built Environment (iiSBE)	Ng et al. (2007); iiSBE (2009); Saraiva et al. (2019)
47	Building Sustainability Assessment Method (BSAM) for Iran	Iran	Shahrzad Malek, David Grierson	Malek and Grierson (2016)
48	Framework for Integrating United Nations Sustainable Development Goals into Sustainable Non-residential Building Assessment and Management in Jordan	Jordan	Rami Alawneh, Farid Ghazali, Hikmat Ali, Ahmad Farhan Sadullah	Alawneh et al. (2019)
49	Environmental and Economic Sustainability Assessment Method for the Retrofitting of Residential Buildings in Turkey	Turkey	Ikbal Cetiner, Ecem Edis	Cetiner and Edis (2014)
50	Conceptual Framework for Implementing Sustainability in the Building Sector	United Kingdom	Peter O. Akadiri, Ezekiel A. Chinyio, Paul O. Olomolaiye	Akadiri et al. (2012)
51	Sustainability Assessment of High School Buildings (SAHSB) in Portugal	Portugal	Tatiana Santos Saraiva, Manuela de Almeida, Luís Bragança	Saraiva et al. (2019)
52	Healthcare Building Sustainability Assessment Method (HBSAtool-PT) for Portugal	Portugal	Maria de Fátima Castro, Ricardo Mateus, Luís Bragança	Castro et al. (2017)

Source: Author's compilation.

References

Akadiri, P. O., Chinyio, E. A., & Olomolaiye, P. O. (2012). Design of a sustainable building: A conceptual framework for implementing sustainability in the building sector. *Buildings*, *2*(2), 126–152.

Akadiri, P. O., & Olomolaiye, P. O. (2012). Development of sustainable assessment criteria for building materials selection. *Engineering, Construction and Architectural Management*, *19*(6), 666–687.

Ali, H. H., & Al Nsairat, S. F. (2009). Developing a green building assessment tool for developing countries – Case of Jordan. *Building and Environment*, *44*(5), 1053–1064.

Berardi, U. (2012). Sustainability assessment in the construction sector: Rating systems and rated buildings. *Sustainable Development*, *20*(6), 411–424.

Bernardi, E., Carlucci, S., Cornaro, C., & Bohne, R. (2017). An analysis of the most adopted rating systems for assessing the environmental impact of buildings. *Sustainability*, *9*(1226), 1–27.

Bhakar, V., Digalwar, A. K., & Sangwan, K. S. (2018). Sustainability assessment framework for manufacturing sector – A conceptual model. *Procedia CIRP*, *69*, 248–253.

Bhatt, R., & Macwan, J. E. M. (2012). Global weights of parameters for sustainable buildings from consultants' perspectives in Indian context. *Journal of Architectural Engineering*, *18*(3), 233–241.

Carvalho, J. P., Bragança, L., & Mateus, R. (2019). Optimising building sustainability assessment using BIM. *Automation in Construction*, *102*, 170–182.

Ching, F. D., & Shapiro, I. M. (2014). *Green building illustrated*. Hoboken, New Jersey: John Wiley & Sons.

Construction Specifications Institute (2013). *The CSI sustainable design and construction practice guide*. Somerset: John Wiley & Sons.

Da Silva, N. A. F. (2015). *Building certification schemes and the quality of indoor environment*. Denmark: Technical University of Denmark.

da Silva, V. G. (2007). Sustainability assessment of buildings: Would LEED lead Brazil anywhere? *CIB world building congress 'Construction for development*, Cape Town, South Africa, 2417–2427.

Ding, G. K. C. (2008). Sustainable construction – the role of environmental assessment tools. *Journal of Environmental Management*, *86*(3), 451–464.

Fowler, K. M., & Rauch, E. M. (2006). *Sustainable building rating systems summary*. Richland, WA: Pacific Northwest National Lab.

Fuerst, F., & McAllister, P. (2011). Eco-labeling in commercial office markets: Do LEED and Energy Star offices obtain multiple premiums? *Ecological Economics*, *70*(6), 1220–1230.

Gasparatos, A., & Scolobig, A. (2012). Choosing the most appropriate sustainability assessment tool. *Ecological Economics*, *80*(1), 1–7.

Goh, C. S., & Rowlinson, S. (2013). The roles of sustainability assessment systems in delivering sustainable construction. *29th Annual ARCOM Conference*, 1363–1371.

Goodhew, S. (2016). *Sustainable construction processes: A resource text*. UK: John Wiley & Sons.

Gu, Z., Wennersten, R., & Assefa, G. (2006). Analysis of the most widely used building environmental assessment methods. *Environmental Sciences*, *3*(3), 175–192.

Gusta, S. (2016). Sustainable construction in Latvia – Opportunities and challenges. *15th International Scientific Conference*, 1291–1299.

Haapio, A., & Viitaniemi, P. (2008). A critical review of building environmental assessment tools. *Environmental Impact Assessment Review*, *28*(7), 469–482.

Hastings, R., & Wall, M. (2007). *Sustainable solar housing: Volume 1 – Strategies and solutions*. UK: Earthscan.

He, B., Luo, T., & Huang, S. (2019). Product sustainability assessment for product life cycle. *Journal of Cleaner Production, 206*, 238–250.

Herda, G., Autio, V., & Lalande, C. (2017). *Building sustainability assessment and benchmarking – An introduction*. Nairobi, Kenya: United Nations Settlements Programme (UN-Habitat).

Hes, D. (2007). Effectiveness of 'green' building rating tools: A review of performance. *International Journal of Environmental, Cultural, Economic and Social Sustainability, 3*(4), 143–152.

Huang, R., & Hsu, W. (2011). Framework development for state-level appraisal indicators of sustainable construction. *Civil Engineering and Environmental Systems, 28*(2), 143–164.

International Organisation for Standardization (2011). *ISO 21929-1 sustainability in building construction – Sustainability indicators*. Switzerland: ISO.

International Organisation for Standardization. (2020). *Sustainable development goals*. Retrieved 01 February 2020 from https://www.iso.org/sdgs.html

Jawali, R., & Fernández-Solís, J. L. (2008). A building sustainability rating index (BSRI) for building construction. *Proceedings of the 8th International Post Graduate Research Conference*, 1–16.

Jayawickrama, T. S. (2014). *Conceptual framework for environmental rating systems for infrastructure projects in Sri Lanka: Application to small hydropower projects*. National University of Singapore.

Kaur, H., & Garg, P. (2019). Urban sustainability assessment tools: A review. *Journal of Cleaner Production, 210*, 146–158.

Kubba, S. (2012). *Handbook of green building design and construction: LEED, BREEAM, and green globes*. USA: Butterworth-Heinemann.

Kumanayake, R., & Luo, H. (2017). Development of an automated tool for buildings' sustainability assessment in early design stage. *Procedia Engineering, 196*, 903–910.

Larsson, N. (2004). (2004). An overview of green building rating and labelling systems. *Paper presented at the Symposium on Green Building Labelling*, 15–21.

Latif, H. H., Gopalakrishnan, B., Nimbarte, A., & Currie, K. (2017). Sustainability index development for manufacturing industry. *Sustainable Energy Technologies and Assessments, 24*, 82–95.

Lee, W. L. (2013). A comprehensive review of metrics of building environmental assessment schemes. *Energy and Buildings, 62*, 403–413.

Li, Y., Chen, X., Wang, X., Xu, Y., & Chen, P. (2017). A review of studies on green building assessment methods by comparative analysis. *Energy and Buildings, 146*, 152–159.

Maheepala, S., Ukwattage, M. I., Madugoda, P. M., & Jayasinghe, G. Y. (2017). Green building rating systems: Present status, challenges and future perspectives. *Proceedings of International Forestry and Environment Symposium, 22*, 101–115.

Mahmoud, S., Zayed, T., & Fahmy, M. (2019). Development of sustainability assessment tool for existing buildings. *Sustainable Cities and Society, 44*, 99–119.

Markelj, J., Kitek Kuzman, M., Grošelj, P., & Zbašnik-Senegačnik, M. (2014). A simplified method for evaluating building sustainability in the early design phase for architects. *Sustainability, 6*(12), 8775–8795.

Ng, S. T., Chen, Y., & Wong, J. M. (2013). Variability of building environmental assessment tools on evaluating carbon emissions. *Environmental Impact Assessment Review, 38*, 131–141.

Organisation for Economic Co-operation and Development (2003). *Environmentally sustainable buildings: Challenges and policies*. Paris, France: OECD Publications.

Pope, J., Annandale, D., & Morrison-Saunders, A. (2004). Conceptualising sustainability assessment. *Environmental Impact Assessment Review, 24*(6), 595–616.

Poveda, C. A., & Lipsett, M. (2011). A review of sustainability assessment and sustainability/environmental rating systems and credit weighting tools. *Journal of Sustainable Development, 4*(6), 36.

Presley, A., & Meade, L. (2010). Benchmarking for sustainability: An application to the sustainable construction industry. *Benchmarking: An International Journal, 17*(3), 435–451.

Reeder, L. (2010). *Guide to green building rating systems: Understanding LEED, green globes, Energy Star, the national green building standard, and more.* Hoboken, New Jersey: John Wiley & Sons.

RSMeans. (2011). *Green building: Project planning and cost estimating* (Third ed.). Hoboken, NJ: John Wiley & Sons.

Saad, M. H., Nazzal, M. A., & Darras, B. M. (2019). A general framework for sustainability assessment of manufacturing processes. *Ecological Indicators, 97*, 211–224.

Said, F. S., & Harputlugil, T. (2019). A research on selecting the green building certification system suitable for Turkey. *GRID-Mimarlık, Planlama Ve Tasarım Dergisi, 2*(1), 25–53.

Said, I., Osman, O., Shafiei, M. W., Rashideh, W. M. A., & Kooi, T. K. (2009). *Modeling of construction firm's sustainability.* Malaysia: ICCI, 1–12.

Sebake, T. N. (2008). Review of appropriateness of international environmental assessment tools for a developing country. *World Sustainable Building Conference*, 1–7.

Sev, A. (2011). A comparative analysis of building environmental assessment tools and suggestions for regional adaptations. *Civil Engineering and Environmental Systems, 28*(3), 231–245.

Siebers, R., Kleist, T., Lakenbrink, S., Bloech, H., den Hollander, J., & Kreißig, J. (2016). *Sustainability certification labels for buildings.* UK: Wiley Online Library.

Trusty, W. B. (2000). Introducing an assessment tool classification system. *Advanced Building Newsletter, 25*(7), 1–2.

Uğur, L. O., & Leblebici, N. (2018). An examination of the LEED green building certification system in terms of construction costs. *Renewable and Sustainable Energy Reviews, 81*, 1476–1483.

Waas, T., Hug, J., Block, T., Wright, T., Benitez-Capistros, F., & Verbruggen, A. (2014). Sustainability assessment and indicators: Tools in a decision-making strategy for sustainable development. *Sustainability, 6*(9), 5512–5534.

Wallhagen, M., Glaumann, M., Eriksson, O., & Westerberg, U. (2013). Framework for detailed comparison of building environmental assessment tools. *Buildings, 3*(1), 39–60.

Weerasinghe, U. G. D. (2012). *Development of a framework to assess sustainability of building projects.* Canada: University of Calgary.

Whang, S., & Kim, S. (2015). Balanced sustainable implementation in the construction industry: The perspective of Korean contractors. *Energy and Buildings, 96*, 76–85.

Yıldız, S., Yılmaz, M., Kıvrak, S., & Gültekin, A. B. (2016). Neighborhood sustainability assessment tools and a comparative analysis of five different assessment tools. *Journal of Planning, 26*(2), 93–100.

Yu, W., Cheng, S., Ho, W., & Chang, Y. (2018). Measuring the sustainability of construction projects throughout their lifecycle: A Taiwan lesson. *Sustainability, 10*(5), 1523.

7   Sustainability Tools and Methods for Building Evaluation

Introduction

Green Building (GB) or sustainable construction (SC) is aimed at encompassing the dimensions of sustainable development (i.e., social, economic, and environmental) in building design, planning, construction, maintenance, and occupation (Gusta, 2016). Characterised by their environmentally friendly features, the risk premium for investors, reduced holdings costs, and additional benefits for users, GBs are now in high demand and used as a marketing tool (Hui et al., 2017). As the importance of sustainable practices and environmental consciousness continues to increase, evaluating the green credentials of buildings has become a crucial aspect of the construction and built environment sector at large. To therefore ascertain how green a building is, specific minimum sustainability requirements must be complied with during evaluation. Several tools, systems, and methods have been established to assess and measure the sustainability performance of buildings to provide standardised benchmarks and frameworks for evaluating their green attributes. These tools, systems, and methods consider various aspects of building design, construction, operation, and their impact on the environment, providing valuable insights into the overall sustainability of a building or infrastructure project. According to Zhang (2015), a green building rating tool (GBRT) must be concise, user-friendly, scientific, and comprehensive to ensure that GBs are practised and wholly accepted by the people. GBRTs reflect the importance of the sustainability concept in buildings and construction works by creating environmental awareness in building practices (Tambovceva et al., 2012).

Existing Tools, Systems, and Methods for Evaluating Green Buildings

The introduction and proliferation of GBRTs are on the increase globally, thereby becoming an important strategy for creating sustainability awareness (Masara, 2019). GBRTs are created to provide a system for rating, measuring, scoring, assessing, or evaluating a building's environmental impact and for promoting the adoption of GBs (Ng et al., 2007). According to Saraiva et al. (2019), these tools have the potential to address environmental issues by minimising construction and demolition waste, reducing water and energy consumption, mitigating the impact

DOI: 10.1201/9781003415961-10

of deforestation, and reducing the heat island effects of buildings. Notable among the existing GBRTs are Building Research Establishment Environmental Assessment Method (BREEAM), Leadership in Energy in Environmental Design (LEED), Comprehensive Assessment System for Built Environment Efficiency (CASBEE), and Green Star. The likes of BREEAM and LEED which originated from the United Kingdom (UK) and the United States of America (USA) respectively have been adapted for use in other countries around the world. However, there has been an increased interest and progress in the development of new GBRTs that are country-specific and market-based to address the peculiar sustainability concerns of each country (Malek & Grierson, 2016). There is also a need to constantly update these tools based on regional/country requirements and sustainability objectives (Mishra & Kauškale, 2018). The following notable and currently used GBRTs are examined to establish an understanding of their constituent evaluation categories and criteria.

Building Research Establishment Environmental Assessment Method

The BREEAM was the world's first building assessment tool created in the year 1990 (Kubba, 2012). BREEAM was co-developed by the Building Research Establishment (BRE) and the Energy and Environment Canada (ECD) for use in the UK (CSI, 2013). It is a voluntary and market-oriented tool for evaluating a building's environmental capability to maintain consistency in objectivity and level of quality (Olgyay & Herdt, 2004). BREEAM is the leading and most widely used sustainability assessment tool as most of the country-specific tools stemmed from it or even adopted it. With over 75 countries implementing BREEAM since its creation in 1990, it remains the most widely used assessment tool owing to its flexibility (Doan et al., 2017).

BREEAM is characterised by its applicability to a range of building types, comprehensive approach, sound research, and global acceptance making it appropriate fundamentals upon which specialised applications are built (Mah, 2011). BREEAM offers developers, designers, clients, and other stakeholders in the industry the following:

- A tool to help minimise running costs, and improve living and working environments;
- A benchmark higher than regulation;
- A standard that shows a progression towards organisational and corporate environmental objectives;
- Market recognition for low environmental impact buildings;
- Assurance that best environmental practice is integrated into a building/project; and
- Inspiration to find innovative solutions that reduce environmental impact (BRE, 2018).

According to Sev (2011), the core objectives of BREEAM are to ensure the best environmental practices in building design, operation, and management; reduce in environmental footprint; and increase awareness of building impacts on the environment. To date, BREEAM has been used to conduct over 530,000 building assessments across a range of building types, namely office, retail, data centres, educational, residential, healthcare, industrial, mixed-use, and other buildings (BRE, 2018). However, a major issue raised by construction professionals and researchers is that BREEAM focusses only on the environment and fails to address the combination of the triple bottom line of sustainability (Abrahams, 2017).

BREEAM Technical Standards

In response to criticisms and evolving agendas in the construction industry, BREEAM has subscribed to regular upgrades in line with the UK. building regulations, taking into consideration more sustainability criteria and project types. From the earliest two versions for assessing homes and offices, there has been the development of other versions to address the various categories of buildings and projects. This has led to the establishment of BREEAM versions for the assessment of industrial, eco-homes, courts, healthcare, offices, prisons, education, multi-residential, communities, and retail buildings/projects (Reeder, 2010; Weerasinghe, 2012). As informed by BRE (2018), the use of BREEAM is at present confined to the following technical standards:

a   BREEAM Communities (new communities or regeneration projects);
b   BREEAM Infrastructure (infrastructure projects);
c   BREEAM New Construction (design, construction, intended use, and future-proofing of new building developments, new homes, and new-build extensions to existing buildings);
d   BREEAM In-Use (all existing commercial type buildings excluding residential dwellings for now);
e   BREEAM Refurbishment and Fit-Out (design and works of a refurbishment or fit-out project such as homes and heritage buildings); and
f   BREEAM Bespoke (Bespoke projects and mixed-use developments).

In this research study, the focus is on BREEAM New Construction (which is more robust and comprehensively encompassing the components of other technical standards) in informing the formulation of the biomimicry sustainability assessment tools (BioSAT) for new constructions (homes and commercial buildings). The BREEAM New Construction technical standard applies to building projects that are residential dwellings, data centres, offices, healthcare, industrial, educational, residential institutions, and public and community buildings, hence it's a choice for the conceptualised tool in this book.

BREEAM Area of Measured Performance

BREEAM areas of measured performance (criteria) are the categories by mean of which it evaluates and addresses sustainable values. The ten categories under which credits are allocated are water, waste, energy, innovation, management, transport, pollution, materials, land use and ecology, health, and wellbeing (Reeder, 2010; Weerasinghe, 2012; CSI, 2013; BRE, 2018). The BREEAM categories predominantly focus on and evaluate the environmental performance of a building from the design, procurement, and construction stages to the management and operational stage with little or no attention paid to the social and economic aspects of sustainability. Table 7.1 presents the aim and issues addressed by each of the categories for compliance.

BREEAM Assessment Rating Benchmarks

To successfully rate a building project, credit weighting systems/scores are crucial. To determine the final rating of the building project which is the main result from a certified BREEAM assessment, weightings, and credits are allocated for each category to provide a total sum (Chen et al., 2015). As indicated by Shukla et al. (2015), BREEAM assessment credit scores are capped at 100% and are classified into the following rating benchmarks as shown in Figure 7.1 and Table 7.2. The building project's final score is calculated to be able to award a BREEAM rating level according to the rating benchmarks. It is, however, stipulated that a minimum performance standard in six key categories out of the ten must be attained before a BREEAM rating performance can be given (Chen, 2017).

BREEAM Assessment Process

The BREEAM assessment process is the step-by-step guide to obtaining a BREEAM rating for a building project or development. According to Gibberd (2003), a BREEAM assessment can take place in two ways. The first way is a mock assessment which is conducted by filling out the prediction checklist to provide an approximate estimation of the performance and rating of the project/development. The estimated result achieved in this mock assessment does not culminate in the final rating for the project/development as it is not submitted to the certification body for a decision. However, the second way is the full and standard assessment process which is undertaken by the appointed licensed BREEAM assessor. For the full and standard assessment process to be carried out, there must be evidence that relates to specific components of the project/development to show compliance with relevant requirements (BREEAM categories). This evidence is in the form of necessary project information and documentation to attest to the conformity of each assessment category to a certain acceptable minimum standard. A summary of the BREEAM certification process as graphically represented by Shukla et al. (2015) is shown in Figure 7.2. This is in alignment with BRE (2018), which highlighted the following laid-down process to obtain a BREEAM rating for any project/development:

a Decide which BREEAM standard (i.e., BREEAM Infrastructure, BREEAM Communities) applies to the development,

b Appoint a licensed BREEAM Assessor to assess the project or building to the correct BREEAM standard,

Table 7.1 BREEAM categories of compliance

S/N	Category	Aim	Issues
1	Energy	Encourages the specification and design of energy-efficient building solutions, systems, and equipment that support the sustainable use of energy in the building and sustainable management of the building's operations. Measures to improve the inherent energy efficiency of the building, encourage the reduction of carbon emissions, and support efficient management throughout the operational phase of the building's life are assessed.	Issues in this category assess measures to improve the inherent energy efficiency of the building, encourage the reduction of carbon emissions, and support efficient management throughout the operational phase of the building's life.
2	Health and Well-being	Encourages the increased comfort, health, and safety of building occupants, visitors, and others within the vicinity. This category aims to enhance the quality of life in buildings by recognising those that encourage a healthy and safe internal and external environment for occupants.	Issues in this category aim to enhance the quality of life in buildings by recognising those that encourage a healthy and safe internal and external environment for occupants.
3	Innovation	Provides opportunities for exemplary performance and innovation (products and processes) to be recognised that are not included within or go beyond the requirements of the credit criteria. This includes exemplary performance credits, for where the building meets the exemplary performance of an issue. Innovative products and processes are also included for which an innovation credit can be claimed.	Issues in this category aim to foster cost-saving benefits of innovation, facilitated by helping encourage, drive, and publicise the accelerated uptake of innovative measures (new technology, process, and practices).
4	Land Use and Ecology	Encourages sustainable land use, habitat protection and creation, and improvement of long-term biodiversity for the building's site and surrounding land.	Issues in this category relate to the reuse of brownfield sites or those of low ecological value, the mitigation and enhancement of ecology, and long-term biodiversity management.
5	Materials	Encourages steps taken to reduce the impact of construction materials through design, construction, maintenance, and repair.	Issues in this category focus on the procurement of materials that are sourced in a responsible way and have a low embodied impact over their life, including environmental impacts (due to extraction, processing and manufacture, and recycling).

(Continued)

Table 7.1 (Continued)

S/N	Category	Aim	Issues
6	Management	Encourages the adoption of sustainable management practices in the design, construction, commissioning, handover, and aftercare activities to ensure that robust sustainability objectives are set and followed through into the building operation.	Issues in this category focus on embedding sustainability actions through the key stages of design, procurement, and initial occupation from the initial project brief stage to the appropriate provision of aftercare.
7	Pollution	Prevents and controls pollution and surface water run-off associated with the building's use and location.	Issues within this category aim to reduce building's impact on surrounding communities and the environment arising from light pollution, noise, flooding and emissions to air, land, and water pollution.
8	Transport	Enables access to sustainable means of transport for building users.	Issues in this category focus on the accessibility of public transport and other alternative transport solutions (cyclist facilities, provision of amenities local to a building) that support reductions in car journeys and, therefore, congestion and CO_2 emissions over the life of the building. It also addresses the design and provision of transport and movement infrastructure to encourage the use of sustainable modes of transport and long-term stewardship of the development.
9	Waste	Promotes sustainable management (and reuse where feasible) of construction, operational waste, and waste through future maintenance and repairs associated with the building structure.	Issues in this category aim to reduce the waste arising from building construction and operation, encouraging its diversion from landfills. It includes recognition of measures to reduce future waste as a result of the need to alter the building in light of future changes to climate.
10	Water	Promotes sustainable use of water in the operation of the building and its site.	Issues in this category focus on identifying means of reducing potable water consumption (internal and external) over the lifetime of the building and minimising loss through leakage.

Source: BRE (2018).

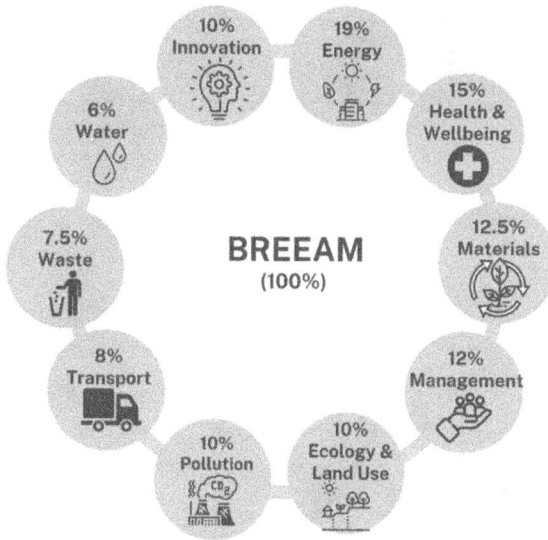

Figure 7.1 BREEAM rating benchmarks

c Register the project for assessment through the appointed licensed BREEAM Assessor,

d Carry out a pre-assessment with the assistance of the licensed BREEAM Assessor utilising their experience and expertise,

e As the project and assessment progress, collate the necessary project information (evidence) and pass this to the licensed BREEAM Assessor,

f The licensed BREEAM Assessor will review the information and determine compliance with BREEAM standards,

g The licensed BREEAM Assessor will submit their assessment report to the certification body for a certification decision (rating), and

h A BREEAM rating certificate for the project/development will be issued and the achievement will be showcased with a case study, BREEAM banner, or plaque from the BREEAM webstore.

Table 7.2 BREEAM rating benchmarks

BREEAM rating	*Percentage score (%)*	*Remarks*
Outstanding	≥ 85	Innovator
Excellent	≥ 70	Best practice
Very Good	≥ 55	Advanced good practice
Good	≥ 45	Intermediate good practice
Pass	≥ 30	Standard good practice
Unclassified	< 30	Non-compliant

Source: Chen et al. (2015), Doan et al. (2017), and BRE (2018).

Figure 7.2 Summary of BREEAM certification process

BREEAM International New Construction 2016 Scheme Version

The BREEAM International (New Construction) 2016 scheme is because of the continuous effort to provide assessment schemes that address the various building types and stages in construction. The aim is to evaluate the environmental lifecycle impacts of new projects at the design and construction stage. As defined by BRE (2017), 'new constructions are new standalone projects/developments, or a new extension to an existing structure which will come into use for the first time upon completion of the works'. The 2016 version of the BREEAM International (New Construction) scheme is meant to provide an assessment of the listed building types in Table 7.3.

BREEAM International New Construction Categories,
Assessment Issues, and Credits

BREEAM International NC assesses the ten categories of energy, management, water, transport, health and wellbeing, materials, waste, pollution, innovation, and land use and ecology, with weightings assigned to them respectively. According to Chen (2017), weightings and credits achieved are incorporated for each category to provide a total assessment score, thereby determining the final grade (rating) of the project/development. Each of the categories in BREEAM International NC is further broken down into subsets to provide a comprehensive understanding and administration of credit ratings. As contained in the BREEAM International NC 2016 technical manual, the ten categories are discussed as follows.

Table 7.3 Building types covered under BREEAM International New Construction scheme

Sector	Building type	Description
Residential	Residential	Single and multiple dwellings
Commercial	Offices	General office buildings, offices with research and development (R&D) areas (category 1 laboratories)
	Industrial	Industrial unit (warehouse storage or distribution) and (process, manufacturing, or vehicle servicing)
	Retails	Shop or shopping centre, retail park or warehouse, showroom, hot food takeaway, 'over the counter' service provider, e.g., financial, estate and employment agencies, and betting offices, restaurant, café and drinking establishment
Education		Preschool, schools and colleges, universities, higher education institutions
Residential Institutions	Long-term stays	Residential care home, sheltered accommodation, residential college or school (halls of residence), local authority secure residential accommodation, military barracks
Hotels and Residential Institutions	Short-term stays	Residential training centre, secure training centre, hotel, hostel, boarding and guest house
Non-standard building types	Bespoke	Prison, law court, fire station, police station, gallery or museum, library, cinema, place of worship, conference facility, community or visitor centre, theatre or concert hall, town hall or civic centre, sports or leisure facility (with or without a pool), hospital and other healthcare facility, transport hub (coach, bus or rail station), research and development (category 2 or 3 laboratories – non-higher education)

Source: BRE (2017).

MANAGEMENT

The management (MAN) category aims to ensure that sustainable management practices are adopted for the design, construction, commissioning, handover, and aftercare stages of the project. By adopting these practices, sustainability objectives are identified and acted upon through the cogent stages of the construction project. The management category consists of five sub-categories, namely project brief and design tagged MAN 01, life cycle cost and service life planning tagged MAN 02, responsible construction practices tagged MAN 03, commissioning and handover tagged MAN 04, and aftercare tagged MAN 05. The MAN 01 assessment sub-category is aimed at recognising and encouraging an integrated design process that optimises building performance. The MAN 02 assessment sub-category aims to deliver whole life value by encouraging the use of life cycle costing to improve design, specification, through-life maintenance, and operation, and through the dissemination of capital cost reporting to promote economic sustainability. The MAN 03 assessment sub-category aims to recognise and encourage construction sites that are managed in an environmentally and socially considerate, responsible, and accountable manner. The MAN 04 assessment sub-category aims to encourage a properly planned handover and commissioning process that reflects the needs of the building occupants. The MAN 05 assessment sub-category aims to provide post-handover aftercare to the building owner or occupants during the first year of occupation to ensure the building operates and adapts, where relevant, following the design intent and operational demands.

HEALTH AND WELLBEING

The health and wellbeing (HEA) category of compliance for the BREEAM rating system encourages the increased safety, comfort, and health of building occupants, visitors, and others within the vicinity. Assessment issues in this category are aimed at enhancing the quality of life in buildings by giving attention to those factors that encourage a safe and healthy environment (internal and external) for occupants. The health and wellbeing category consists of nine sub-categories, namely visual comfort (HEA 01), indoor air quality (HEA 02), safe containment in laboratories (HEA 03), thermal comfort (HEA 04), acoustic performance (HEA 05), accessibility (HEA 06), hazards (HEA 07), private space (HEA08), and water quality (HEA 09).

The HEA 01 assessment sub-category aims to ensure daylighting, artificial lighting, and occupant controls are considered at the design stage to ensure best practices in visual performance and comfort for building occupants. The HEA 02 assessment sub-category aims to recognise and encourage a healthy internal environment through the specification and installation of appropriate ventilation, equipment, and finishes. The HEA 03 assessment sub-category aims to recognise and encourage a healthy internal environment through the safe containment and removal of pollutants. The HEA 04 assessment sub-category aims to ensure that appropriate thermal comfort levels are achieved through design, and controls are selected to maintain a thermally comfortable environment for occupants within the building. The HEA 05 assessment sub-category aims to ensure the building's acoustic performance, including sound insulation meets the appropriate standards for its purpose. The HEA 06 assessment sub-category aims to recognise and encourage effective measures that promote safe and secure use, and access to and from the building. The HEA 07 assessment sub-category is aimed at reducing or negating the impact of a natural hazard on the building. The HEA 08 assessment sub-category is aimed at providing an external space that gives occupants privacy and a sense of wellbeing. The HEA 08 assessment sub-category aims to minimise the risk of water contamination in building services and ensure the provision of clean, fresh sources of water for building users.

ENERGY

The energy (ENE) category of compliance for the BREEAM rating system encourages the design and specification of energy-efficient building systems that aid the sustainable use and management of energy in the building and the building's operation. This category evaluates ways of improving the inherent energy efficiency of the building, encourages the reduction of carbon footprint and emissions, and supports efficient management throughout the operational phase of the building's life cycle. The energy category consists of nine sub-categories, namely reduction of energy use and carbon emissions (ENE 01), energy monitoring (residential and non-residential) – ENE 02, external lighting (ENE 03), low carbon design (ENE 04), energy efficient cold storage (ENE 05), energy-efficient transport systems (ENE 06), energy-efficient laboratory systems (ENE 07), energy-efficient equipment (ENE 08), and drying space (ENE 09).

The ENE 01 assessment sub-category aims to recognise and encourage buildings designed to minimise operational energy demand, primary energy consumption, and CO emissions. The ENE 02a assessment sub-category aims to recognise and encourage the installation of energy sub-metering that facilitates the monitoring of operational energy consumption. The ENE 02b assessment sub-category aims to recognise and encourage monitoring of energy consumption using energy display devices. The ENE 03 assessment sub-category aims to recognise and encourage the specification of energy-efficient light fittings for external areas of project development. The ENE 04 assessment sub-category aims to encourage the adoption of design measures, which reduce building energy consumption and associated carbon emissions and minimise reliance on active building services systems. The ENE 05 assessment sub-category aims to recognise and encourage the installation of energy-efficient refrigeration systems, thereby reducing operational greenhouse gas emissions resulting from the system's energy use. The ENE 06 assessment sub-category aims to recognise and encourage the specification of energy-efficient transport systems. The ENE 07 assessment sub-category aims to recognise and encourage laboratory areas that are designed to be energy efficient and minimise the CO_2 emissions associated with their operational energy consumption. The ENE 08 assessment sub-category aims to recognise and encourage the procurement of energy-efficient equipment to ensure optimum performance and energy savings in operation. The ENE 09 assessment sub-category aimed at providing a reduced energy means of drying clothes.

TRANSPORT

The transport (TRA) category of compliance for the BREEAM rating system encourages efficient and effective access to sustainable transportation mediums for building users. This category focuses on the accessibility to public transport and other alternative transport solutions that support reductions in car journeys, congestion, and CO emissions over the building's life cycle. The transport category consists of six sub-categories, namely public transport accessibility (TRA 01), proximity to amenities (TRA 02), alternative modes of transport (non-residential, and residential institutions only and residential only) – TRA 03, maximum car parking capacity (TRA 04), travel plan (TRA 05), and home office (TRA 06).

The TRA 01 assessment sub-category aims to recognise and encourage development in proximity of good public transport networks, thereby helping to reduce transport-related pollution and congestion. The TRA 02 assessment sub-category aims to encourage and reward a building location that facilitates easy access to local services and reduces the environmental, social, and economic impacts resulting from multiple or extended building user journeys, including transport-related emissions and traffic congestion. Both TRA 03a and TRA 03b assessment sub-categories aim to provide facilities that encourage building users to travel using low-carbon modes of transport and to minimise individual journeys. The TRA 04 assessment sub-category aims to encourage the use of alternative means of transport other than the private car to and from the building, thereby helping to reduce transport-related emissions and traffic congestion associated with the building's operation. The TRA 05 assessment sub-category aims to recognise the consideration given to accommodate a range of travel options for building users, thereby encouraging the Reduction of reliance on forms of travel that have the highest environmental impact. Lastly, the TRA 06 assessment sub-category aims to reduce the need to commute to work by providing residents with the necessary space and services to be able to work from home.

WATER

The water (WAT) category of compliance for the BREEAM rating system encourages sustainable utilisation of water on site and building operations. This category focuses on identifying the ways of reducing the consumption (external and internal) of potable water and minimising losses through leakage throughout a building's lifetime. The water category consists of four sub-categories, namely water consumption (WAT 01), water monitoring (WAT 02), water leak detection and prevention (WAT 03), and water efficient equipment (WAT 04). The WAT 01 assessment sub-category aims to reduce the consumption of potable water for sanitary use in new buildings from all sources using water-efficient components and water recycling systems. The WAT 02 assessment sub-category aims to ensure water consumption can be monitored and managed, and therefore encourage reductions in consumption. The WAT 03 assessment sub-category aims to reduce the impact of water leaks that may otherwise go undetected. The WAT 04 assessment sub-category aims to reduce water consumption by encouraging the specification of water-efficient equipment.

MATERIALS

The material (MAT) category of compliance for the BREEAM rating system encourages the following processes that reduce the impact of construction materials generated through design, construction, maintenance, and repair. This category focuses on procuring materials that are responsibly sourced and possess a low embodied impact during extraction, processing, manufacturing, and recycling over their life cycle. The material category consists of six sub-categories namely life cycle impacts (MAT 01), hard landscaping and boundary protection (MAT 02), responsible sourcing of construction products (MAT 03), insulation (MAT 04), design for durability and resilience (MAT 05), and material efficiency (MAT 06). The MAT 01 assessment sub-category aims to recognise and encourage the use of robust and appropriate life cycle assessment tools and consequently the specification of construction materials with a low environmental impact (including embodied carbon) over the full life cycle of the building. The MAT 02 assessment sub-category is not assessed as a standalone in this version. The MAT 03 assessment sub-category aims to recognise and encourage the specification and procurement of responsibly sourced construction products. The MAT 04 assessment sub-category is not assessed as a standalone in this version. The MAT 05 assessment sub-category aims to recognise and encourage adequate protection of exposed elements of the building and landscape, thereby minimising the frequency of replacement and maximising materials optimisation. Finally, the MAT 06 assessment sub-category is aimed at recognising and encouraging measures to optimise material efficiency to minimise the environmental impact of material use and waste without compromising on structural stability, durability, or service life of the building.

WASTE

The waste (WST) category of compliance for the BREEAM rating system encourages sustainable management and reuse, where feasible, of construction and operational waste and waste through future maintenance and repairs associated with the building structure. Through good design and construction practices, this category focuses on reducing the waste arising from building construction and operation and encourages its diversion from landfills. Measures to reduce future waste are recognised because of the necessity to alter the building in anticipation of future changes to the climate. The waste category consists of six sub-categories, namely construction waste management (WST 01), recycled aggregates (WST 02), operational waste (non-residential and residential institutions and residential only) – WST 03, speculative finishes (WST 04), adaptation to climate (WST 05), and functional adaptability (WST 06).

The WST 01 assessment sub-category aims to promote resource efficiency via the effective and appropriate management of construction waste. The WST 02 assessment sub-category aims to recognise and encourage the use of recycled and secondary aggregates, thereby reducing the demand for virgin material and optimising material efficiency in construction. The WST 03a (first part of operational waste) assessment sub-category aims to recognise and encourage the provision of dedicated storage facilities for a building's operational-related recyclable waste streams so that this waste is diverted from landfills or incineration. The WST 03b (second part of operational waste) assessment sub-category aims to recognise and encourage the provision of dedicated storage facilities for operational-related household waste streams and so help to avoid waste being sent to landfill or incineration. The WST 04 assessment sub-category aims to encourage the specification and fitting of finishes selected by the building occupant and therefore avoid unnecessary waste of materials. The WST 05 assessment sub-category aims to recognise and encourage measures taken to mitigate the impact of extreme weather conditions arising from climate change over the lifespan of the building. The WST 06 assessment sub-category aims to recognise and encourage measures taken to accommodate future changes in the use of the building over its lifespan.

LAND USE AND ECOLOGY

The land use and ecology (LE) category of compliance for the BREEAM rating system encourages the sustainable use of land, habitat creation and protection, and improvement of long-term biodiversity for the building's site and surrounding land. Reuse of brownfield sites or those of less ecological value, the mitigation and enhancement of ecology, and long-term biodiversity management are the issues that relate to this category. The land use and ecology category consists of five sub-categories, namely site selection (LE 01), the ecological value of the site and protection of ecological features (LE 02), minimising impact on existing site ecology (LE 03), enhancing site ecology (LE 04), and long-term impact on biodiversity (LE 05). The LE 01 assessment sub-category aims to encourage the use of previously occupied or contaminated land and avoid land which has not been previously disturbed. The LE 02 assessment sub-category aims to encourage development on land that already has limited value to wildlife and to protect existing ecological features from substantial damage during site preparation and completion of construction works. The aim of the LE 03 assessment sub-category does not apply to the version addressed. The LE 04 assessment sub-category aims to encourage actions taken to enhance the ecological value of the site because of development. The LE 05 assessment sub-category is aimed at minimising the long-term impact of the development on the site and the surrounding area's biodiversity.

POLLUTION

The pollution (POL) category of compliance for the BREEAM rating system aims at addressing the control and prevention of pollution and surface water run-off associated with the location and use of the building. Reducing the building's effect on neighbouring communities and environments arising from light pollution, flooding, noise, and emissions to air, land and water are the issues addressed by this category. The pollution category consists of five sub-categories namely impact on refrigerants (POL 01), NOx emissions (POL 02), surface water run-off (POL 03), reduction of night-time light pollution (POL 04), and reduction of noise pollution (POL 05). The POL 01 assessment sub-category is aimed at reducing the level of greenhouse gas emissions arising from the leakage of refrigerants used to heat or cool the building. The POL 02 assessment sub-category is aimed at contributing to a reduction in national NOx emission levels using low-emission heat sources in the building. The POL 03 assessment sub-category aims to avoid, reduce, and delay the discharge of rainfall to public sewers and watercourses, thereby minimising the risk and impact of localised flooding on and off-site, watercourse pollution, and other environmental damage. The POL 04 assessment sub-category aims to ensure that external lighting is concentrated in the appropriate areas and that upward lighting is minimised, reducing unnecessary obtrusive light pollution, energy consumption, and nuisance to neighbouring properties. The POL 05 assessment sub-category aims to reduce the likelihood of noise arising from fixed installations on the new development affecting nearby noise-sensitive buildings.

INNOVATION

The innovation (INN) category of compliance for the BREEAM rating system provides opportunities for recognising exemplary performance and innovation that are not included within or go beyond the requirements of the credit criteria. This includes innovative products and processes for which one innovation credit can be claimed as approved by BRE Global Ltd and exemplary performance credits (where the building meets the exemplary performance levels of an issue). By encouraging, driving, and publicising the accelerated uptake of innovative measures, the cost-saving benefits of innovation are facilitated and fostered. The innovation category provides up to a maximum of 10 credits, with the total BREEAM score capped at 100%. The credit score is assigned when the building demonstrates exemplary performance by meeting one or more of the following assessment issues, namely responsible construction practices (MAN 03), aftercare (MAN 05), indoor air quality (HEA 02), reduction of energy use and carbon emissions (ENE 01), alternative modes of transport (TRA 03a and TRA 03b), water consumption (WAT 01), life cycle impacts (MAT 01), responsible sourcing of construction products (MAT 03), construction waste management (WST 01), recycled aggregates (WST 02) and adaptation to climate change (WST 05). The 'Innovation' assessment sub-category is aimed at supporting innovation within the construction industry through the recognition of sustainability-related benefits which are not rewarded by standard BREEAM issues.

Leadership in Energy in Environmental Design (LEED)

The LEED rating system was developed in the year 2000 by the United States Green Building Council (USGBC) as a voluntary consensus-based standard (Suzer, 2015; Wu et al., 2016). Since its establishment, LEED has become one of the most widely used assessment tools (Zarghami et al., 2018), especially in North America (US and Canada) with its acceptability and reputation for credibility globally. As a point-based system, the objective of LEED is to ensure energy efficiency and a significant reduction in the negative environmental impacts of building projects (Uğur & Leblebici, 2018). The LEED rating system is a compilation of sustainability performance standards utilised in the certification of different building types in private and public spaces to promote durable, healthy, and environmentally sound practices (Kubba, 2016). To date, LEED has a significant presence in over 165 countries and territories with a high record of more than 90,000 projects (USGBC, 2019). LEED-rated buildings, especially those that scored 'excellent' for the energy performance category are known to belong to the top 5% in the GB market (Lee & Burnett, 2008).

The LEED GB rating system is available for almost all building and community project types, providing a framework to create efficient, cost-saving, and healthy buildings and environments. The LEED rating system, as the name implies, evaluates primarily the environmental aspect of a building for the sustainability triangle while placing much emphasis on the health and well-being of occupants (Siebers et al., 2016). The following seven objectives were emphasised by the USGBC to guide the successful deployment and application of LEED rating systems:

  i Establishing a more ecological economy;
 ii Reversing the contribution to climate change;
iii Protecting and restoring water resources;
 iv Promoting sustainable and regenerative material loops;
  v Improving the health and wellbeing of individuals;
 vi Improving social justice, environmental awareness, and quality of life; and
vii Protecting and restoring the diversity of species and ecosystems (Goodhew, 2016; Siebers et al., 2016).

As one of the leading GB rating systems, a building project that is LEED-rated is assumed to be energy-efficient (Scofield, 2009). The assertion that LEED projects are energy-saving is justified by the encouraging results of early studies on LEED buildings (Scofield, 2013). Using the points scheme that allocates credits for building categories that aim at improving sustainability (reductions in energy use and improvements in indoor environment quality), the LEED rating system encourages an integrated design approach (Newsham et al., 2009). LEED provides building projects with instant recognition; higher resale value; healthier indoor and outdoor space for building occupants, the community, and the environment; faster lease-up rates; lower use of water, energy, and other resources; and classification as a leader in GB (USGBC, 2019). LEED is also adjudged to be one of the few rating systems that are easy to use owing to its user-friendly interface and product support (Nguyen & Altan, 2011). The LEED rating system is administered by the Green Business Certification Institute (GBCI) by performing verification and third-party technical reviews of LEED-registered building projects.

LEED Technical Standards

In a changing world, LEED has subscribed to continuous improvement with each newer version raising the sustainability bar above the previous version. LEED versions v1, LEED 1.0 pilot, v2 LEED 2.0, v2 LEED 2.1, v2 LEED 2.2, v3 LEED 2008, v3 LEED 2009, v4 LEED v4, and the latest LEED v4.1 are all true reflections of concerted effort towards ensuring the LEED rating system achieves the supreme target of holistic sustainability. Version 4.1 is the newest LEED rating system which is a direct result of lessons learned from LEED users and efforts aimed at arriving at solutions that address different markets. The latest version of the LEED rating system (LEED v4.1) is a holistic upgrade of previous versions and it addresses broader categories of buildings and projects at different developmental stages. The classification of LEED rating systems is based on parameters such as building, construction, and market (Rastogi et al., 2017). Among these project types are schools, hospitality, data centres, warehouses and distribution centres, and healthcare facilities among others (USGBC, 2019; Wu et al., 2019). According to the USGBC (2019), the application of LEED v4.1 is presently confined to the following:

LEED v4.1 BUILDING DESIGN AND CONSTRUCTION (BD+C)

The LEED v4.1 BD+C addresses new construction and major renovations (used for buildings that do not primarily serve warehouses and distribution centres, residential, data centres, k-12 educational, retail and hospitality or healthcare uses); schools (used for buildings that constitute ancillary and core learning spaces on k-12 school premises, higher education and non-academic buildings on school premises); core and shell development (used for new construction or major renovation of buildings for the exterior shell and core plumbing, mechanical and electrical units and appropriate where more than 40% of the gross floor area is incomplete as at the time of certification); data centres (used for buildings designed and equipped to meet the needs of voluminous computing equipment used for data processing and storage); retail (used for buildings that are for the conduct of retail sale of consumer product goods); healthcare (used for hospital buildings that operate 24 hours a day and seven days a week in-patient medical treatment, including acute and long-term care); hospitality (used for buildings dedicated to inns, motels, hotels and other businesses within the hospitality industry providing short-term or transitional lodging without or with food); and warehouses and distribution centres (used for buildings that are for storing goods, manufactured products, raw materials or personal belongings such as self-storage).

LEED v4.1 INTERIOR DESIGN AND CONSTRUCTION (ID+C)

LEED v4.1 ID+C is the rating system for interior spaces that are a complete interior fit-out where at least 60% of the gross floor area of the project must be complete at the certification time. It addresses retail (used for interior spaces for conducting retail sales of consumer product goods and including showroom and preparation/storage areas that support customer service); commercial interiors (used for interior spaces addressing functions other than retail or hospitality); and hospitality (used for buildings dedicated to inns, motels, hotels and other businesses within the hospitality industry providing short-term or transitional lodging without or with food).

LEED v4.1 BUILDING OPERATIONS AND MAINTENANCE (O+M)

LEED v4.1 O+M is the rating system for existing buildings undergoing rehabilitation/renovation works with little to no construction. For such a building to be categorised as 'existing', it must be occupied and fully operational for at least one year. The entire building's gross floor area must also be included in the project. LEED v4.1 O+M rating system is specifically for existing interior spaces (spaces serving hospitality, retail or commercial) that are contained within a portion of an existing building (existing interiors) and existing whole buildings (existing buildings).

LEED v4.1 RESIDENTIAL – BUILDING DESIGN AND CONSTRUCTION (BD+C)

LEED v4.1 Residential is the required rating system for all single and multi-family residential projects which are either new constructions or major renovations. The project must include the entire building's gross floor area where at least 60% of the project's gross floor area must be completed by the time of certification. LEED v4.1 Residential addresses new family homes that are detached or attached, multifamily buildings with up to four units (single-family homes); multifamily buildings with two or more units and any number of storeys (multifamily homes); and new construction or renovation of multifamily buildings (multifamily homes core and shell).

LEED v4.1 CITIES AND COMMUNITIES

The LEED v4.1 for Cities and Communities is a novel way forward for achieving green, smart, inclusive, and resilient cities by providing a globally consistent method of measuring and communicating performance. According to project types, it addresses new cities and communities that are in the designing/planning stage (cities in the planning and design stage), and cities and communities that are not more than 75% built out (cities in the existing stage).

LEED v4.1 RECERTIFICATION

LEED v4.1 Recertification aims to protect building assets by helping to improve and maintain the building while preserving the sustainability investment. This rating system is available to all in-use and occupied projects that were previously LEED-certified. LEED encourages projects to upgrade themselves by monitoring and reporting their performance data to prove that their building projects are performing as expected. By doing so, such projects qualify for recertification with the latest version (LEED v4.1) which will be valid for three years. This book focuses on LEED v4.1 BD+C and addresses new construction and major renovations (which also encompass the components of other technical standards) as part of the GB tools examined in the formulation of the conceptualised sustainability assessment tool for new constructions (homes and commercial buildings).

LEED Areas of Measured Performance

According to Rastogi et al. (2017), the main objective of sustainability assessment tools/systems is to evaluate the sustainability performance of building projects. To meet this objective, LEED as one of the major sustainability assessment tools has constantly subscribed to continuous revisions, corrections, and upgrades of the various versions offered. By introducing and adding more stringent and explicit sustainability parameters and credit scoring mechanisms, each latest version of the LEED rating system is a step closer to holistically achieving sustainability. However, LEED rating systems still tend strongly towards addressing the environmental dimension of the sustainability pillar.

LEED areas of measured performance (also known as criteria) are the categories according to which a building project is evaluated to determine its sustainability performance. These criteria serve as yardsticks and a way of regulating and guiding the CI and its activities towards achieving a sustainable future (Thilakaratne & Lew, 2011). The nine categories are the themes under which a building project is evaluated for sustainability with the possibility of scoring points. These categories are Integrative Process-IP, Sustainable Sites-SS, Location and Transportation-LT, Water Efficiency-WE, Energy and Atmosphere-EA, Materials and Resources-MR, Indoor Environmental Quality-IEQ, Innovation-IN and Regional Priority-RP (Goodhew, 2016; Siebers et al., 2016; Wu et al., 2016; Ding et al., 2018). Table 7.4 presents the evaluation categories addressed by LEED version 4.1 BD+C (New Construction).

Table 7.4 LEED v4.1 BD+C New Construction evaluation categories

Assessment category	Assessment criteria	Credits available	Total credits (110)
Integrative Process (IP)	Integrative Process	1	1
Location and Transportation (LT)	Sensitive land protection	1	16
	High priority site	2	
	Surrounding density and diverse uses	5	
	Access to quality transit	5	
	Bicycle facilities	1	
	Reduced parking footprint	1	
	Electric vehicles	1	
Sustainable Sites (SS)	Construction activity pollution prevention	Required	10
	Site assessment	1	
	Protect or restore habitat	2	
	Open space	1	
	Rainwater management	3	
	Heat island reduction	2	
	Light pollution reduction	1	
Water Efficiency (WE)	Outdoor water use reduction	Required	11
	Indoor water use reduction	Required	
	Building-level water metering	Required	
	Outdoor water use reduction	2	
	Indoor water use reduction	6	
	Cooling tower water use	2	
	Water metering	1	
Energy and Atmosphere (EA)	Fundamental commissioning and verification	Required	33
	Minimum energy performance	Required	
	Building-level energy metering	Required	
	Fundamental refrigerant management	Required	
	Enhanced commissioning	6	
	Optimise energy performance	18	
	Advanced energy metering	1	
	Grid harmonisation	2	
	Renewable energy	5	
	Enhanced refrigerant management	1	

(Continued)

Table 7.4 (Continued)

Assesment category	Assessment criteria	Cresdits available	Total credits (110)
Materials and Resources (MR)	Storage and collection of recyclables	Required	13
	Construction and demolition waste management planning	Required	
	Building life-cycle impact reduction	5	
	Building product disclosure and optimisation – environmental product declarations	2	
	Building product disclosure and optimisation – sourcing of raw materials	2	
	Building product disclosure and optimisation – material ingredients	2	
	Construction and demolition waste management	2	
Indoor Environmental Quality (IEQ)	Minimum indoor air quality performance	Required	16
	Environmental tobacco smoke control	Required	
	Enhanced indoor air quality strategies	2	
	Low-emitting materials	3	
	Construction indoor air quality management plan	1	
	Indoor air quality assessment	2	
	Thermal comfort	1	
	Interior lighting	2	
	Daylight	3	
	Quality views	1	
	Acoustic performance	1	
Innovation (IN)	Innovation	5	6
	LEED accredited professional	1	
Regional Priority (RP)	Regional priority: specific credit	4	4

Source: USGBC (2019).

LEED Rating System Benchmarks

The continually evolving simple point-based system of LEED has made it the most popular and widely accepted GBRT in the US (Kubba, 2016; Awadh, 2017). A building project is evaluated against the seven LEED categories by allocating credit points. A total of 100 base points is available per LEED rating system while the allocation of ten bonus points for innovation and regional evaluation categories brings the total achievable points to 110, although a minimum of 40 points is required for certification. For 80+ points, the 'Platinum' level of certification is awarded, 60 to 79 points are awarded 'Gold', 50 to 59 are awarded 'Silver', and 40 to 49 points are awarded 'Certified' (Awadh, 2017; Turk et al., 2018).

Comprehensive Assessment System for Building Environment Efficiency (CASBEE)

Globally, Japan is one of the pioneering countries in the development of GB rating tools when it established the CASBEE for evaluating the sustainability performance of building projects (Sharifi et al., 2012). CASBEE was developed in the year 2001 by a research committee (academia, industry, and local and national government) of the Japan Sustainable Building Consortium (JSBC) and the Institute for Building Environment and Energy Conservation (IBEC) to reduce the adverse environmental impacts associated with the built environment (Sasatani et al., 2015; JSBC & IBEC, 2020). As the most studied GBRT after LEED and BREEAM (Li et al., 2017), CASBEE remains the most notable and well-known rating system in Japan with international prominence. According to the study of Wong and Abe (2014), CASBEE is applied mainly for the following purposes in Japan: a building design tool (by architects and engineers); a sustainable building reporting system (by local governments); GB certification for marketing purposes (by developers and building owners); an environmental labelling tool for asset assessment (by developers and building owners); a basis for judging competitive design proposals (by developers and building owners); and a basis for eligibility for preferential loan interest rates (by financing institutions).

According to Murakami et al. (2010) and Shwe et al. (2017), CASBEE family tools (CASBEE-Housing, CASBEE-Building, CASBEE-Urban Development, CASBEE-City) are tailored to address different scales such as city management, urban (town development), and construction (housing and buildings). CASBEE tools for building scale include CASBEE for New Construction (CASBEE-NC), CASBEE for Existing Buildings (CASBEE-EB), CASBEE for Renovation (CASBEE-RN), CASBEE for Interior Space, CASBEE for Temporary Construction (CASBEE-TC), CASBEE for Heat Island Relaxation (CASBEE-HI), CASBEE for Schools, and CASBEE for Real Estate. The CASBEE assessment process differs from other GBRTs as it employs an additive/weighting system different from the conventional process of adding points obtained in all evaluation categories (Sev, 2011). A building project can be awarded either S (excellent), A (very good), B+ (good), B- (rather poor), or C (poor) based on the total scores achieved (Kibert, 2016), which is based on a scale of 1 to 5 with 3 as the standard minimum score.

CASBEE assessment works by evaluating buildings based on six sustainability issues that are further categorised into two sections, namely ***Environmental Quality of the Building (Q)*** which contains Indoor Environment (Q1), Quality of Service (Q2) and Outdoor Environment On-site (Q3), and ***Environmental Load Reduction of the Building (LR)*** which contains Energy (LR1), Resources and Materials (LR2) and Off-site Environment (LR3). According to the JSBC (2014) and

Shamseldin (2018), the assessment result, known as the Built Environment Efficiency (BEE) result, is calculated from the scores for the environmental quality of the building (SQ) and scores for environmental load reduction of the building (SLR) based on the formula below:

$$BEE = \frac{Q: \text{Environmental Quality of the Building}}{L: \text{Environmental Load Reduction of the Building}} = \frac{25 \times (SQ - 1)}{25 \times (5 - SLR)}$$

Table 7.5 shows the evaluation categories, criteria, and sub-criteria for CASBEE-NC while Table 7.6 presents a detailed interpretation of the assessment results and their corresponding certification awarded.

Table 7.5 CASBEE-NC 2014 edition rating system categories

Assessment category	Assessment criteria	Sub-criteria
Indoor Environment (Q1)	Sound Environment	Noise
		Sound insulation
		Sound absorption
	Thermal Comfort	Room temperature control
		Humidity control
		Type of air conditioning system
	Lighting and Illumination	Daylight
		Anti-glare measures
		Illuminance level
		Lighting controllability
	Air Quality	Source control
		Ventilation
		Operation plan
Quality of Service (Q2)	Service Ability	Functionality and usability
		Amenity
		Maintenance
	Durability and Reliability	Earthquake resistance
		Service life of components
		Reliability
	Flexibility and Adaptability	Spatial margin
		Floor load margin
		System renewability
Outdoor Environment (Q3)	Preservation and Creation of Biotope	n/a
	Townscape and Landscape	n/a
	Local Characteristics and Outdoor Amenity	Attention to local character and improvement of comfort
		Improvement of the thermal environment on-site

(Continued)

Table 7.5 (Continued)

Assesment category	Assessment criteria	Sub-criteria
Energy (LR1)	Control of Heat Load on the Outer Surface of Buildings	n/a
	Natural Energy Utilisation	n/a
	Efficiency in Building Service System	n/a
	Efficient Operation	Monitoring
		Operation and management system
Resources and Materials (LR2)	Water Resources	Water saving
		Rainwater and grey water
	Reducing Use of Non-renewable Resources	Reducing use of materials
		Continuing use of existing structural frame, etc.
		Use of recycled materials as structural materials
		Use of recycled materials as non-structural materials
		Timber from sustainable forestry
		Efforts to enhance the reusability of components and materials
	Avoiding the Use of Materials with Pollutant Content	Use of materials without harmful substances
		Elimination of CFCs and halons
Off-site Environment (LR3)	Consideration of Global Warming	n/a
	Consideration of Local Environment	Air pollution
		Heat island effect
		Load on local infrastructure
	Consideration of Surrounding Environment	Noise, vibration, and odour
		Wind damage and daylight obstruction
		Light pollution

Source: Kawazu et al. (2005) and JSBC (2014).

Table 7.6 CASBEE-NC 2014 edition rating system benchmarks

Ranks	Assessment	BEE value and details	Expression
S	Excellent	BEE = 3.0 or more and Q = 50 or more	5-Stars
A	Very Good	BEE = 1.5 to 3.0	4-Stars
		BEE = 3.0 or more and Q is less than 50	
B +	Good	BEE = 1.0 to 1.5	3-Stars
B –	Fairly Poor	BEE = 0.5 to 1.0	2-Stars
C	Poor	BEE = less than 0.5	1-Star

Source: JSBC (2014).

Deutsche Gesellschaft Für Nachhaltiges Bauen (DGNB) Certification System

To ensure the practical application of GBs, the Deutsche Gesellschaft für Nach-haltiges Bauen (DGNB), which is the German Sustainable Building Council (GSBC) developed its certification system in 2009. The certification system which is a plan-ning, design, and optimisation tool for assessing GB projects is based on the concept of sustainability, placing balanced prominence on commercial viability, people, and the environment (Eberl, 2010; GSBC, 2020). The DGNB certification system is used in Denmark, Austria, and Switzerland and the certification is administered by the Green Building Council Denmark (DK-GBC), the Austrian Sustainable Building Council (ÖGNI), and the Swiss Sustainable Building Council (SGNI) respectively.

The DGNB certification system consists of schemes for the following build-ing types: existing buildings, new construction, interiors, and districts (DGNB, 2020). Based on a uniform evaluation method that considers all aspects of GB, the certification system addresses six evaluation categories, namely Environmental Quality (ENV), Economic Quality (ECO), Sociocultural and Functional Quality (SOC), Technical Quality (TEC), Process Quality (PRO) and Site Quality (SITE). Table 7.7 presents the evaluation categories and criteria of the DGNB certification system. Based on the weighting of each category assessed environmental quality, economic quality, and sociocultural and functional quality account for 22.5% each, technical quality account for 15%, process quality account for 12.5%, and site quality account for 5%, total performance index is calculated before certification is awarded. Table 7.8 presents the total performance index and its corresponding certification awards for the DGNB system.

Table 7.7 DGNB System 2018 Version New Buildings evaluation categories

Assessment category	Assessment category group	Assessment criteria
Environmental Quality (ENV)	Effects on the Global and Local Environment (ENV1)	Building life cycle assessment
		Local environmental impact
		Sustainable resource extraction
	Resource Consumption and Waste Generation (ENV2)	Potable water demand and wastewater volume
		Land use
		Biodiversity at the site
Economic Quality (ECO)	Life Cycle Costs (ECO1)	Life cycle cost
	Economic Development (ECO2)	Flexibility and adaptability
		Commercial viability
Sociocultural and Functional Quality (SOC)	Health, Comfort, and User Satisfaction (SOC1)	Thermal comfort
		Indoor air quality
		Acoustic comfort
		Visual comfort
		User control
		Quality of indoor and outdoor spaces
		Safety and security
	Functionality (SOC2)	Design for all

(Continued)

Table 7.7 (Continued)

Assesment category	Assessment category group	Assessment criteria
Technical Quality (TEC)	Technical Quality (TEC1)	Sound insulation
		Quality of the building envelope
		Use and integration of building technology
		Ease of cleaning building components
		Ease of recovery and recycling
		Emissions control
		Mobility infrastructure
Process Quality (PRO)	Planning Quality (PRO1)	Comprehensive project brief
		Sustainability aspects in tender phase
		Documentation for sustainable management
		Urban planning and design procedure
	Construction Quality Assurance (PRO2)	Construction site/construction process
		Quality assurance of the construction
		Systematic commissioning
		User communication
		FM-compliant planning
Site Quality (SITE)	Site Quality (SITE1)	Local environment
		Influence on the district
		Transport access
		Access to amenities

Source: DGNB (2018).

Living Building Certification (LBC)

The Living Building Challenge (LBC) was developed by the Cascadia Green Building Institute and the International Living Future Institute (ILFI), formerly known as the International Living Building Institute (Thiel et al., 2013). The LBC, which is a certification tool by the ILFI, aims to restore the relationship between nature and people and to realign the global ecological footprint to be within the planet's carrying capacity (ILFI, 2019). However, a comparative analysis by Zimmermann et al. (2019) revealed that the LBC certification system focuses more on the social and environmental aspects of sustainability and less consideration for the social aspect. The Living Certification and Petal Certification are the two certifications under the LBC that can be applied to four building project scopes (i.e.,

Table 7.8 DGNB certification system benchmarks

Total performance index	Minimum performance index	Awards expression
From 35%	n/a	Bronze
From 50%	35%	Silver
From 65%	50%	Gold
From 80%	65%	Platinum

Source: DGNB (2018).

new buildings, existing buildings, interior projects, and landscape or infrastructure projects). Despite the claim to be the most comprehensive certification system in terms of sustainability standards, LBC does not use a scoring system and is less specific regarding the means of adjudicating compliance with each evaluation category (Patenaude & Plouffe, 2015). The LBC consists of seven performance areas (known as Petals) for evaluation, namely place, water, energy, health and happiness, materials, equity, and beauty (ILFI, 2020). Table 7.9 presents the evaluation categories and criteria of the LBC certification system.

Building Environmental Assessment Method (BEAM) Plus

The Building Environmental Assessment Method (BEAM) Plus which was formerly the Hong Kong Building Environmental Assessment Method (HK-BEAM) established in 1996, is a voluntary certification scheme developed and administered by the BEAM Society Limited (Cheng & Venkataraman, 2012; Adegbile, 2013). While BEAM Plus is owned and operated by BEAM Society Limited, the Hong Kong Green Building Council (HKGBC) certifies BEAM Plus projects, and accredits BEAM Assessors (BAS), BEAM Affiliate (BA), and BEAM Professionals (BEAM Pro). BEAM assessment and certification are endorsed by the Hong Kong Government as they present stakeholders with a single sustainability performance label of building projects aimed at achieving a more efficient, functional, comfortable, healthier, and safer built environment (BEAM Society Limited, 2012).

Table 7.9 LBC Version 4.0 Certification System evaluation categories

Assessment category/Petal	Assessment criteria/Imperative
Place	Ecology of place
	Urban agriculture
	Habitat exchange
	Human scaled living
Water	Responsible water use
	Net positive water
Energy	Energy + carbon reduction
	Net positive energy
Health + Happiness	Healthy interior environment
	Healthy interior performance
	Access to nature
Materials	Responsible materials
	Red list
	Responsible sourcing
	Living economy sourcing
	Net positive waste
Equity	Universal access
	Inclusion
Beauty	Beauty + biophilia
	Education _ inspiration

Source: ILFI (2019).

According to Lee and Burnett (2008), when BEAM Plus was first launched, provisions were made for two versions to address new buildings (HK-BEAM 1/96) and existing office buildings (HK-BEAM 2/96). By the best international practice, the BEAM Society has subscribed to continuously update and upgrade the assessment tool to meet current national and international sustainability demands. A revised version of the former HK-BEAM was launched in 2009 with a series of updates to date (Hui et al., 2017). While BEAM Plus certification significantly addresses the economic and environmental aspects of SD, little attention is given to the socio-cultural dimension (Adegbile, 2013). The BEAM Plus New Buildings version 2.0 (NB v2.0) released in 2019 contains seven evaluation categories for which credits are assigned, namely Integrated Design and Construction Management (IDCM), Sustainable Sites (SS), Materials and Waste (MW), Energy Use (EU), Water Use (WU), Health and Wellbeing (HWB), and Innovations and Additions (IA). Table 7.10 presents the evaluation categories, criteria, credits, and weightings of the BEAM Plus NB v2.0 while Table 7.11 shows the grading system for awarding certification.

Comprehensive Environmental Performance Assessment Scheme for Buildings

Under the 2001 Government Policy Objectives, Ove Arup and Partners Hong Kong Ltd was commissioned by the Buildings Department of the Government of Hong Kong Special Administrative Region (HKSAR) to undertake a consultancy study for the initiation of the Comprehensive Environmental Performance Assessment Scheme (CEPAS) for GB evaluation (Wu & Yau, 2005). CEPAS is an award-based evaluation tool aimed at encouraging improvements in the design, construction/demolition, and operation of buildings. There are CEPAS assessment manuals for the pre-design, design, construction, and operation stages of a building project. Eight sustainability performance categories are evaluated in CEPAS, namely Indoor Environmental Quality (IE), Building Amenities (BA), Resources Use (RE), Loadings (LD), Site Amenities (SA), Neighbourhood Amenities (NA), Site Impacts (SI), and Neighbourhood Impacts (NI). The evaluation categories, criteria, sub-criteria, and weightings of CEPAS are presented in Table 7.12.

Green Star Australia

Launched in 2003 by the Green Building Council of Australia, Green Star is a voluntary and the only national rating tool aimed at improving the environmental efficiency, health, and well-being of buildings and communities in Australia (GBCA, 2020). The rating tool delivers independent validation of sustainable results throughout the built environment life cycle (GBCA, 2017). Green Star-Communities (precinct planning and development), Green Star-Design and As-built (building design and construction), Green Star-Interiors (fit-out design and construction), and Green Star-Performance (Building operations and maintenance) are the four rating tools available within Green Star to assess various building projects. The Green Star rating tools evaluate a wide range of sustainability issues that tend towards the environmental aspect with points awarded to calculate the overall certification award (Xia et al., 2013).

Table 7.10 BEAM Plus NB v2.0 evaluation categories

Assessment category	Assessment criteria	Credits available	Total credits	Category weightings
Integrated Design and Construction Management (IDCM)	Sustainability champions – project	Required	25+14 bonus	18%
	Environmental management plan	Required		
	Timber used for temporary works	Required		
	Sustainability champions – design	1+1 bonus		
	Complimentary certification	3 bonus		
	Integrated design process	4		
	Life cycle costing	1		
	Commissioning	4		
	Sustainability champions – construction	1		
	Measures to reduce site emissions	4		
	Construction and demolition waste recycling	2+4 bonus		
	Construction IAQ management	1		
	Considerate construction	1		
	Building management manuals	1		
	Operator training plus chemical storage room	1		
	Digital facility management interface	1 bonus		
	Occupant engagement platform	1 bonus		
	Document management system	2		
	BIM integration	1+1 bonus +2 bonus		
	Design for engagement and education on green buildings	1+1 bonus		

(Continued)

Table 7.10 (Continued)

Assessment category	Assessment criteria	Credits available	Total credits	Category weightings
Sustainable Sites (SS)	Minimum landscaping requirements	Required	20+19 bonus	15%z
	Pedestrian-oriented and low-carbon transport	2+1 bonus +2 bonus		
	Neighbourhood amenities	2		
	Building design for sustainable urbanism	2+1 bonus +1 bonus		
	Neighbourhood daylight access	1		
	Noise control for building equipment	1		
	Light pollution control	2		
	Biodiversity	1+2 bonus +3 bonus		
	Urban heat island mitigation	For site area < 1000m2: 1 For Site area ≥ 1000m2: 4 + 2 bonus + 4 bonus		
	Immediate neighbourhood wind environment	1		
	Outdoor thermal comfort	2		
	Stormwater management	2 +1 bonus		
	Design for climate change adaptation	1 bonus +1 bonus		
Materials and Waste (MW)	Minimum waste handling facilities	Required	14+21 bonus	9%
	Building re-use	2 bonus +1 bonus		
	Modular and standardised design	1+1 bonus		
	Prefabrication	1+3 bonus		
	Design for durability and resilience	1+2 bonus		
	Sustainable forest products	1+1 bonus		
	Recycled materials	1+2 bonus		
	Ozone-depleting substances	2		
	Regional materials	1+2 bonus		
	Use of certified green products	2+3 bonus +1 bonus		
	Life cycle assessment	1		
	Adaptability and deconstruction	1+1 bonus		
	Enhanced waste-handling facilities	2+2 bonus		

(Continued)

Table 7.10 (Continued)

Assessment category	Assessment criteria	Credits available	Total credits	Category weightings
Energy Use (EU)	Minimum energy performance	Required	31+13 bonus	29%
	Low carbon passive design	6		
	Reduction of CO_2 emissions	10+5 bonus		
	Peak electricity demand reduction	3		
	Metering and monitoring	1+2 bonus		
	Renewable and alternative energy systems	6+5 bonus		
	Air-conditioning units	2		
	Clothes drying facilities	1+1 bonus		
	Energy-efficient appliances	2		
Water Use (WU)	Minimum water-saving performance	Required	12+3 bonus	7%
	Annual water use	3+1 bonus		
	Water efficient irrigation	2+1 bonus		
	Water efficient appliances	1		
	Water leakage detection	1		
	Twin tank system	1		
	Cooling tower water	1		
	Effluent discharge to foul sewers	1		
	Water harvesting and recycling	2+1 bonus		

(*Continued*)

Table 7.10 (Continued)

Assessment category	Assessment criteria	Credits available	Total credits	Category weightings
Health and Wellbeing (HWB)	Minimum ventilation performance	Required	19+10 bonus	22%
	Healthy and active living	1 bonus		
	Biophilic design	1 bonus +1 bonus		
	Inclusive design	1+1 bonus		
	Enhanced ventilation	3+1 bonus		
	Waste odour control	1		
	Acoustics and noise	4+1 bonus		
	Indoor vibration	1		
	Indoor air quality	4+1 bonus		
	Thermal comfort	2+1 bonus		
	Artificial lighting	2		
	Daylight	2 bonus		
	Biological contaminations	1		
Innovations and Additions (IA)	Innovations and additions		Maximum 10 bonus	n/a

Source: BEAM Society Limited (2019).

Table 7.11 BEAM Plus NB v2.0 grading system

Grade	Minimum percentage for each category	Total score
Platinum	20%	≥ 75
Gold	20%	≥ 65
Silver	20%	≥ 55
Bronze	20%	< 40

Source: BEAM Society Limited (2019).

Table 7.12 CEPAS for buildings 2006 edition construction stage evaluation categories

Assessment category	Assessment criteria	Sub-criteria
Indoor Environment Quality (IE)	Health and Hygiene	Health and hygiene
	Indoor Air Quality	Indoor air quality strategies
	Lighting Environment	Visual quality and comfort
	Air Quality	Source control
		Ventilation
		Operation plan
Building Amenities (BA)	Safety	n/a
	Management	Building management
Resources Use (RE)	Energy Efficiency	Energy efficiency
	Water Conservation	Water conservation strategies
	Timber Use	Timber for temporary use
		Minimisation of timber use
	Material Use	Recycled material use
		Construction waste recycling
		Demolition waste recycling
		Environmentally friendly materials
Loadings (LD)	Pollution	n/a
		Air pollution
		Water pollution
		Noise pollution
	Waste Management	n/a
		C&D waste management
Site Amenities (SA)	Landscape	Tree preservation
	Security	Security
Neighbourhood Amenities (NA)	Environmental Economics	Environmental economics
Site Impacts (SI)	Nature Conservation	Nature conservation
	Heritage Conservation	Heritage conservation
	Buildability	Buildability
Neighbourhood Impacts (NI)	Environmental Impact Assessment	n/a
	Environmental Interactions	Environmental nuisance
	Impacts to Communities	Impacts to communities

Source: Building Department HKSAR (2006).

Green Star evaluates nine sustainability performance categories namely Management, Indoor Environment Quality, Energy, Transport, Water, Materials, Land Use and Ecology, Emissions, and Innovation (Mitchell, 2010) which contain 30 criteria. These categories account for 100 credit points with an additional 10 points available under the 'Innovation' category, making a total of 110. According to Gandhi and Jupp (2014), certification can be awarded based on the percentage thresholds of the Green Star rating scale. Table 7.13 presents the evaluation categories, criteria, and weightings of Green Star Design and As Built version 1.2.

Green Mark Scheme

The Building and Construction Authority (BCA) Green Mark Scheme was launched in 2005 to drive Singapore's CI towards a more sustainable path (BCA, 2017). The Green Mark Scheme is a voluntary tool that assesses the overall environmental performance

Table 7.13 Green Star Design & As Built v1.2 evaluation categories

Assessment category	Assessment criteria	Credit point	Available points
Management	Green Star Accredited Professional	1	14
	Commissioning and Tuning	4	
	Adaptation and Resilience	2	
	Building Information	1	
	Commitment to Performance	2	
	Metering and Monitoring	1	
	Responsible Construction Practices	2	
	Operational Waste	1	
Indoor Environment Quality	Indoor Air Quality	4	17
	Acoustic Comfort	3	
	Lighting Comfort	3	
	Visual Comfort	3	
	Indoor Pollutants	2	
	Thermal Comfort	2	
Energy	Greenhouse Gas Emissions	20	22
	Peak Electricity Demand Reduction	2	
Transport	Sustainable Transport	10	10
Water	Potable Water	12	12
Materials	Life Cycle Impacts	7	14
	Responsible Building Materials	3	
	Sustainable Products	3	
	Construction and Demolition Waste	1	
Land Use and Ecology	Ecological Value	3	6
	Sustainable Sites	2	
	Heat Island Effect	1	
Emissions	Stormwater	2	5
	Light Pollution	1	
	Microbial Control	1	
	Refrigerant Impacts	1	
Innovation	Innovation	10	10

Source: GBCA (2017).

and impact of buildings by providing a detailed framework in areas such as energy efficiency (Government of Singapore, 2019). Within the Green Mark Scheme, there are rating tools aimed at evaluating new buildings (non-residential, residential, data centres, and landed housing), existing buildings (non-residential, residential, data centres, and schools), user-centric (office interior, retail, supermarket, restaurant, and laboratories), and beyond buildings (districts, parks, and infrastructure). The Green Mark Non-Residential Building (NRB) 2015 is structured into 5 sections consisting of 16 categories and 52 indicators. Green Mark Residential Building (RB) 2016 also contains 16 categories and 34 indicators structured into 5 sections. Points are calculated based on these assessment indicators before certification is awarded based on the total points achieved. Table 7.14 presents the assessment sections, categories, criteria, and corresponding credit points for Green Mark RB:2016.

Building for Ecologically Responsive Design Excellence (BERDE) Program

Initiated and administered by the Philippine Green Building Council (PHILGBC), the Building for Ecologically Responsive Design Excellence (BERDE) certification programme was designed to evaluate the sustainability performance of building projects in the Philippines (ESCI, 2020). It was established in 2009 as a response to the need of the Philippine construction industry to address the adverse impacts of climate change (PHILGBC, 2020). BERDE is recognised as the national voluntary GB rating system by the Department of Energy (DOE) of the Philippine government (PHILGBC, 2015). The BERDE GB rating system is structured into 9 evaluation categories and 29 indicators/criteria. The nine sustainability performance categories evaluated in BERDE are Management, Use of Land and Ecology, Energy Efficiency and Conservation, Water Efficiency and Conservation, Waste Management, Green Materials, Transportation, Indoor Environment Quality, and Emissions. Based on the outcomes of the assessment and total weighting achieved, certification is awarded. The assessment categories and indicators are shown in Table 7.15.

Greenship

Based on the consideration for regulations/standards, natural character, and conditions that apply in Indonesia, the Green Building Council Indonesia (GBC Indonesia) developed and issued the GREENSHIP rating tool to evaluate buildings towards achieving sustainability (GBC Indonesia, 2020). To address different building project categories, five tools are developed within GREENSHIP, namely GREENSHIP New Building, GREENSHIP Existing Building, GREENSHIP Interior Space, GREENSHIP Homes, and GREENSHIP Neighbourhood. Six sustainability categories are evaluated under the GREENSHIP rating tools with each category consisting of criteria for which credit points are allocated to determine the assessment result. The evaluation categories are Appropriate Site Development (ASD), Energy Efficiency and Conservation (EEC), Water Conservation (WAC), Material Resources and Cycle (MRC), Indoor Health and Comfort (IHC), and Building Environment Management (BEM). Table 7.16 presents the assessment categories and criteria for GREENSHIP New Building version 1.1.

Table 7.14 Green Mark RB: 2016 evaluation categories

Assessment section	Assessment categories	Assessment criteria	Credit points	Total points
Climatic Responsive Design	Leadership	Climatic and contextual responsive brief	1	35
		Integrated design process	2	
		Environmental credentials of project team	2	
		Building information modelling	2	
		User engagement	1	
	Urban Harmony	Sustainable urbanism	5	
		Integrated landscape and waterscape	5	
	Tropicality	Tropical façade performance	5	
		Internal organisation	2	
		Ventilation performance	10	
Building Energy Performance	Energy Efficiency	Air conditioning	6	25
		Lighting efficiency	4	
		Car park energy	2	
	Energy Effectiveness	Energy-efficient practices, design, and features	5	
	Renewable Energy	Feasibility study	0.5	
		Solar ready roof	1.5	
		Replacement energy	6	
Resource Stewardship	Water	Water efficiency measures	9	35
		Water usage monitoring	1	
		Alternative water sources	3	
	Materials	Sustainable construction	8	
		Embodied energy	2	
		Sustainable products	8	
	Waste	Environmental construction management plan	1	
		Operational waste management	3	
Smart and Healthy Building	Indoor Air Quality	Occupant comfort	2	25
		Contaminants	6	
	Spatial Quality	Lighting	5	
		Acoustics	2	
		Wellbeing	2	
	Smart Building Operations	Energy monitoring	2	
		Demand control	2	
		Integration and analytics	2	
		System handover and documentation	2	
Advanced Green Efforts	Enhanced Performance	n/a	15	20
	Demonstrating Cost Effective Design	n/a	2	
	Complementary Certifications	n/a	1	
	Social Benefits	n/a	2	

Source: BCA (2017).

Table 7.15 BERDE Green Building Rating System evaluation categories

Assessment category	Assessment criteria
Management (MN)	Green Building Professional
	Stakeholder Consultation
	Project Management
	Certified Green Building
Use of Land and Ecology (LE)	Land Reuse
	Ecological Features Improvement
	Vegetated Open Spaces Promotion
	Heat Island Reduction
Energy Efficiency and Conservation (EN)	Energy Consumption Reduction
	Renewable Energy
Water Efficiency and Conservation (WT)	Water Consumption Reduction
	Graywater Reuse
	Rainwater Collection
Waste Management (WS)	Waste Management
Green Materials (MT)	Green Procurement
	Local Materials
Transportation (TR)	Key Establishments' Proximity
	Mass Transportation Access
	Preferred Parking
	Cyclists and Pedestrians Amenities
Indoor Environment Quality (EQ)	Visual Comfort
	Daylight and Outdoor View Access
	Thermal Comfort
	Acoustic Comfort
	Indoor Air Quality
	Microbial Contamination Prevention
	Low VOC Environment
Emissions (EM)	Greenhouse Gas Inventory
	Refrigerants

Source: PHILGBC (2020).

Green Standard for Energy and Environmental Design (G-SEED)

Between the years 1997 and 2000, South Korea recorded the first GB effort towards the development of a tool to evaluate residential and office buildings (Bahaudin et al., 2014). Based on the adaptation of the GBTool, this effort gave birth to the Green Building Certification Criteria (GBCC) which were deployed by the South Korea Institute of Energy Research (KIER) in 2001 (Bahaudin et al., 2012). The study of Yeom and Lee (2015) which was on the improvement of the Green Standard for Energy and Environmental Design (G-SEED) certification (formerly GBCC) aligns with the global clamour for the continuous update of GBRTs following local or national sustainability demands.

G-SEED certification aims to optimise the sustainability performance of the South Korean built environment by evaluating and certifying building projects in a bid to conserve resources and reduce CO_2 emissions and other negative environmental impacts (Lee & Shepley, 2019). G-SEED certification parameters differ according to the building types to be assessed (Kim & Park, 2018). There are

Table 7.16 GREENSHIP New Building Version 1.1 evaluation categories

Assessment category	Assessment criteria
Appropriate Site Development (ASD)	Basic Green Area
	Site Selection
	Community Accessibility
	Public Transportation
	Bicycle
	Site Landscaping
	Micro Climate
	Storm Water Management
Energy Efficiency and Conservation (EEC)	Electrical Sub Metering
	OTTV Calculation
	Energy Efficiency Measure
	Natural Lighting
	Ventilation
	Climate Change Impact
	On-Site Renewable Energy (Bonus)
Water Conservation (WAC)	Water Metering
	Water Calculation
	Water Use Reduction
	Water Fixtures
	Water Recycling
	Alternative Water Resource
	Rainwater Harvesting
	Water Efficiency Landscaping
Material Resources and Cycle (MRC)	Fundamental Refrigerant
	Building and Material Reuse
	Environmentally Friendly Material
	Non-ODS Usage
	Certified Wood
	Prefab Material
	Regional Material
Indoor Health and Comfort (IHC)	Outdoor Air Introduction
	CO_2 Monitoring
	Environmental Tobacco Smoke Control
	Chemical Pollutants
	Outside View
	Visual Comfort
	Thermal Comfort
	Acoustic Level
Building Environment Management (BEM)	Basic Waste Management
	GP as a Member of The Project Team
	Pollution of Construction Activity
	Advanced Waste Management
	Proper Commissioning
	Submission Green Building Data
	Fit Out Agreement
	Occupant Survey

Source: Green Building Indonesia (2012).

different versions of G-SEED certification for the evaluation of projects such as lodging facilities, retail markets, schools, multi-residential and detached housing, office buildings, and multi-family housing units (Tae & Shin, 2009; Mok & Cho, 2014). An update of G-SEED certification was done in 2016 and it evaluates sustainability under eight categories, namely Indoor Environment, Ecology, Maintenance, Water Management, Materials and Resources, Energy and Environmental Pollution, Land Use and Transportation, and Innovative Design (Yun et al., 2018). Table 7.17 presents the assessment categories and criteria for the New Residential version of G-SEED certification.

Table 7.17 G-SEED certification categories for New Residential

Assessment category	Assessment criteria	Credits available
Land Use and Transportation	Ecological value of existing land	2
	Avoiding excessive underground development	3
	Minimisation of earthwork volume	2
	Validity of countermeasures against solar light interference	2
	Creation of pedestrian roads in the complex and connection with roads for pedestrians	2
	Proximity of public transportation	2
	Suitability of bicycle parking lot and bicycle road	2
	Accessibility of living convenience facilities	1
Energy and Environmental Pollution	Energy performance	12
	Energy monitoring and management support device	2
	Use of new and renewable energy	3
	Application of low-carbon energy source technology	1
	Prohibition of the use of certain substances to protect the ozone layer	2
Materials and Resources	Use of the Environmental Declaration Product (EPD)	4
	Use of low-carbon materials	2
	Use of resource recycling materials	2
	Use of materials for reducing hazardous substances	2
	Application rate of green building materials	4
	Installation of storage facilities for recyclable resources	1
Water Circulation Management	Rainwater management	5
	Use of rainwater and runoff groundwater	4
	Use of water-saving equipment	3
	Water usage monitoring	2
Maintenance	Environmental management plan at construction site	2
	Operation and maintenance documents and manuals	2
	User Manual	2
	Provision of information related to green building certification	3
Ecological Environment	Creation of linked green axis	2
	Natural soil greening rate	4
	Ecological Area Ratio	10
	Biotope composition	4

(*Continued*)

Table 7.17 (Continued)

Assesment category	Assessment criteria		Credits available
Indoor Environment	Application of low-emission products for indoor air pollutants		6
	Secure natural ventilation performance		2
	Securing ventilation performance of unit households		2
	Level of thermostat installation		1
	Lightweight impact sound-blocking performance		2
	Weight impact sound blocking performance		2
	Sound insulation performance of boundary walls between generations		2
	Noise levels indoor and outdoor for traffic noise (road, railway)		2
	Toilet drainage noise		2
Housing Performance Field	Durability		-
	Variability		-
	Considering social weakness of unit households		-
	Social weakness in public spaces		-
	Community centre and facility space		-
	Rate of sunshine in households		-
	Home network integrated system		-
	Security content		-
	Detection and alarm equipment		-
	Smoke control equipment		-
	Fire resistance		-
	Horizontal evacuation distance		-
	Effective width of hallways and stairs		-
	Evacuation equipment		-
	Parts for repairability		-
	Repairability common part		-
Innovative Design	Land use and transportation	Installation of alternative transportation-related facilities	1
	Energy and environmental pollution	Zero energy building	3
		Outer thermal bridge prevention	1
	Materials and resources	Building life cycle evaluation	2
		Reuse of main structural parts of existing buildings	5
	Water circulation management	Reuse of sewage and wastewater treatment water	1
	Maintenance	Environmental management of green construction sites	1
	Ecological environment	Topsoil recycling rate	1
	Green architecture expert	Design participation by green architects	1
	Innovative green architecture planning and design	Evaluate through green building plan and design deliberation	3

Source: G-SEED (2020).

Pearl Rating System for Estidama

According to Qadir et al. (2019), the Pearl Rating System (PRS), which is the first of its kind in the Middle East region is a cardinal part of the Estidama (*Estidama* is the Arabic word for 'sustainability') drive to transform the Abu Dhabi province of the United Arab Emirates (UAE) into a model of sustainable urbanisation. Established in the year 2010 by the Abu Dhabi Urban Planning Council (ADUPC), PRS was developed by incorporating and adapting the elements of BREEAM and LEED to the national environment and unique local needs of the Emirate (Awadh, 2017). The rating tool aims to form a scheme for evaluating sustainability performance beyond the conventional planning and construction stages of buildings (SAS International, 2020). PRS as a government initiative requires all new building development to attain specific standards to minimise environmental impacts by recycling building materials and reducing construction waste, water, and energy use amongst others (Assaf & Nour, 2015).

The PRS is based on four dimensions, also known as the 'four pillars of Estidama' which are social, cultural, economic, and environmental, and comprise two types of credit rating, namely required and optional (AbdelAzim et al., 2017). Required credits (R) are mandatory points that must be attained by every building project submitted for assessment while optional credits are voluntary for which certain points can be achieved. The PRS for Estidama currently comprises the following four tools:

a   Pearl Community Rating System (PCRS): planning;
b   Public Realm Rating System (PRRS): design and construction;
c   Pearl Building Rating System (PBRS): design and construction; and
d   Pearl Villa Rating System (PVRS): design and construction.

To help achieve the Estidama (sustainability) agenda of Abu Dhabi, PRSs are deployed to assess building projects against seven categories of integrated development processes, natural systems, liveable buildings, precious water, resourceful energy, stewarding materials, and innovating practice (Alobaidi et al., 2015). The PBRS which is one of the core aspects of the PRS for Estidama and the reference for this research study, addresses the following building types, their sites, and associated facilities, namely general, office, retail, school, multi-residential, and mixed-use building projects (ADUPC, 2010). Projects that fulfil the requirement of all mandatory credits are rated '1 Pearl', those with all mandatory credits plus 60 credit points are rated '2 Pearl', those with all mandatory credits plus 85 credit points are rated '3 Pearl', those with all mandatory credits plus 115 credit points are rated '4 Pearl', and those with all mandatory credits plus 140 credit points are rated '5 Pearl'. Table 7.18 presents the assessment categories and criteria for the PBRS.

Green Rating for Integrated Habitat Assessment (GRIHA)

The Energy and Resources Institute (TERI) developed the Green Rating for Integrated Habitat Assessment (GRIHA) which was later adopted in 2007 by

Table 7.18 Pearl Building Rating System categories

Assessment category	Assessment criteria	Credits available	Maximum credits
Integrated Development Process (IDP)	Integrated development strategy	R	13
	Tenant fit-out design & construction guide	R	
	Basic commissioning	R	
	Life cycle costing	4	
	Guest worker accommodation	2	
	Construction environmental management	2	
	Building envelope verification	1	
	Re-commissioning	2	
	Sustainability communication	2	
Natural Systems (NS)	Natural systems assessment	R	12
	Natural systems protection	R	
	Natural systems design and management strategy	R	
	Reuse of land	2	
	Remediation of contaminated land	2	
	Ecological enhancement	2	
	Habitat creation & restoration	6	
Liveable Buildings (LB) Outdoors	Plan 2030	R	37
	Urban systems assessment	R	
	Outdoor thermal comfort strategy	R	
	Improved outdoor thermal comfort	2	
	Pearl rated communities	1	
	Accessible community facilities	1	
	Active urban environments	1	
	Private outdoor space	1	
	Public transport	3	
	Bicycle facilities	2	
	Preferred car parking spaces	1	
	Travel plan	1	
	Light pollution reduction	1	
Indoors	Healthy ventilation delivery	R	
	Smoking control	R	
	Legionella prevention	R	
	Ventilation quality	3	
	Material emissions: adhesives and sealants	1	
	Material emissions: paints and coatings	1	
	Material emissions: carpet and hard flooring	1	
	Material emissions: ceiling systems	1	
	Material emissions: formaldehyde reduction	2	
	Construction indoor air quality management	2	
	Car park air quality management	1	
	Thermal comfort and controls: thermal zoning	1	
	Thermal comfort and controls: occupant control	2	
	Thermal comfort and controls: thermal comfort modelling	2	
	High-frequency lighting	1	
	Daylight and glare	2	
	Views	1	
	Indoor noise pollution	1	
	Safe and secure environment	1	

(*Continued*)

Table 7.18 (Continued)

Assesment category	Assessment criteria	Credit available	Maximum credits
Precious Water (PW)	Minimum interior water use reduction	R	43
	Exterior water monitoring	R	
	Improved interior water use reduction	15	
	Exterior water use reduction: landscaping	8	
	Exterior water use reduction: heat rejection	8	
	Exterior water use reduction: water features	4	
	Water monitoring and leak detection	4	
	Stormwater	4	
Resourceful Energy (RE)	Minimum energy performance	R	44
	Energy monitoring & reporting	R	
	Ozone impacts of refrigerants and fire suppression systems	R	
	Improved energy performance	15	
	Cool building strategies	6	
	Energy-efficient appliances	3	
	Vertical transportation	3	
	Peak load reduction	4	
	Renewable energy	9	
	Global warming impacts of refrigerants and fire suppression systems	4	
Stewarding Materials (SM)	Hazardous materials elimination	R	28
	Basic construction waste management	R	
	Basic operational waste management	R	
	Non-polluting materials	3	
	Design for materials reduction	1	
	Design for flexibility and adaptability	1	
	Design for disassembly	1	
	Modular flooring systems	1	
	Design for durability	1	
	Building reuse	2	
	Material reuse	1	
	Regional materials	2	
	Recycled materials	6	
	Rapidly renewable materials	1	
	Reused or certified timber	2	
	Improved construction waste management	2	
	Improved operational waste management	2	
	Organic waste management	2	
Innovating Practice (IP)	Innovative cultural and regional practices	1	3
	Innovating practice	2	

Source: UPC (2010).

the Government of India as the national rating system for GBs (GRIHA, 2019). According to Vij (2010), the formulation process of GRIHA began in 2003/2004 when a team of experts from TERI conducted a detailed study on international and notable tools such as BREEAM, LEED, CASBEE, and HK-BEAM to prepare a rating system adaptable for India. The development of GRIHA also considers the provisions of the National Building Code (NBC) of 2005, the Energy

Conservation Building Code (ECBC) of 2007 by the Indian Bureau of Energy Efficiency (BEE), and other local codes, laws, and standards (Vyas et al., 2014). GRIHA was developed to evaluate and rate residential, institutional, and commercial buildings in India with tailored evaluation categories to address regional climatic conditions, national environmental concerns, and indigenous solutions (Sande & Phadtare, 2015). To determine the sustainability performance of a building using GRIHA, points are awarded by experts based on a set of evaluation categories and a rating based on the overall output result (Bansal et al., 2015). Based on the total credit points obtained, a percentile threshold of 25 to 40 is awarded 1 star, 41 to 55 is awarded 2 stars, 56 to 70 is awarded 3 stars, 71 to 85 is awarded 4 stars while the 86 and above threshold is awarded 5 stars. Since its development and deployment for use, GRIHA has been constantly updated to keep up with the growing sustainability demands of India. The 2019 version of GRIHA consists of 11 categories which are subdivided into 30 criteria. Table 7.19 presents the categories, criteria, and corresponding credits of the GRIHA 2019 version.

Indian Green Building Council (IGBC) Rating Systems

The growth of GB has been driven by the Indian Green Building Council (IGBC) since 2001. To keep up with the global sustainability practice, IGBC launched the IGBC Green Building Rating Systems which are tools to facilitate the adoption and implementation of GB practices in India (IGBC, 2014). As informed by the IGBC (2015), the IGBC Green Building Rating Systems consist of the following tools or technical standards: IGBC Green New Buildings, IGBC Green Existing Buildings, IGBC Green Homes, IGBC Green Residential Societies, IGBC Green Interiors, IGBC Green Healthcare, IGBC Green Data Centre, IGBC Green Factory Buildings, IGBC Green Schools, IGBC Green Campus, IGBC Green Cities, IGBC Green Existing Cities, IGBC Green Villages, IGBC Green Townships, IGBC Green SEZs, IGBC Green Landscapes, IGBC Green Mass Rapid Transit System, IGBC Green Existing Mass Rapid Transit System, IGBC Green Resorts, IGBC Green Railway Stations, IGBC Green Affordable Housing, IGBC Green Health and Well-being, and IGBC Green Service Buildings. To encourage and promote the adoption of these tools, the IGBC secured several incentives from the state and central government bodies and agencies in India.

According to Srivastava et al. (2017), the IGBC Green Building Rating Systems addresses evaluates, and awards credits under the following seven categories, namely Sustainable Architecture and Design, Site Selection and Planning, Water Conservation, Energy Efficiency, Building Materials and Resources, Indoor Environmental Quality, and Innovation and Development. The categories, indicators, and corresponding credits for the IGBC Green New Buildings Rating System are presented in Table 7.20. Based on the total credits earned, the following levels of certification and recognition are awarded to the building project under assessment: certified, silver, gold, platinum, and super platinum.

Table 7.19 GRIHA version 2019 evaluation categories

Assessment category	Assessment criteria	Credits available	Total credits
Sustainable Site Planning (SSP)	Green infrastructure	5	100
	Low-impact design strategies	5	
	Designed to mitigate UHIE	2	
Construction Management (CM)	Air and soil pollution control	1	
	Topsoil preservation	1	
	Construction management practices	2	
Energy Optimisation (EO)	Energy optimisation	12	
	Renewable energy utilisation	5	
	Low ODP and GWP materials	1	
Occupant Comfort (OC)	Visual comfort	4	
	Thermal and acoustic comfort	2	
	Maintaining good IAQ	6	
Water Management (WM)	Water demand reduction	3	
	Wastewater treatment	3	
	Rainwater management	5	
	Water quality and self-sufficiency	5	
Solid Waste Management (SWM)	Waste management – post-occupancy	4	
	Organic waste treatment on-site	2	
Sustainable Building Materials (SBM)	Utilisation of alternative materials in buildings	5	
	Reduction in GWP through lifecycle assessment	5	
	Alternative materials for external site development	2	
Life Cycle Costing (LCC)	Lifecycle cost analysis	5	
Socio-Economic Strategies (SES)	Safety and sanitation for construction workers	1	
	Universal accessibility	2	
	Dedicated facilities for service staff	2	
	Positive social impact	3	
Performance Metering and Monitoring (PMM)	Commissioning for final rating	7	
	Smart metering and monitoring	0	
	Operation and maintenance protocol	0	
Innovation (I)	Innovation	5	5

Source: GRIHA Council and TERI (2019).

GREEN^SL Rating System

The GREEN^SL rating system is developed by the Green Building Council of Sri Lanka (GBCSL) based on an expert study of GBRTs used in Singapore, Malaysia, Indonesia, China, and India (GBCSL, 2020). The rating system with provisions for new and existing building types/categories is a set of sustainability performance benchmarks to evaluate the operations and maintenance of public and private buildings in Sri Lanka. The GREEN^SL rating system aims to promote durable, healthy, high-performance, environmentally sound, and affordable practices in Sri Lanka building projects. The rating system addresses, evaluates, and awards credits under

Table 7.20 IGBC Green New Buildings rating system evaluation categories

Assessment category	Assessment criteria	Credits available	Total credits
Sustainable Architecture and Design (SA)	Integrated design approach	1	5
	Site preservation	2	
	Passive architecture	2	
Site Selection and Planning (SSP)	Local building regulations	Required	14
	Soil erosion control	Required	
	Basic amenities	1	
	Proximity to public transport	1	
	Low-emitting vehicles	1	
	Natural topography or vegetation	2	
	Preservation or transplantation of trees	1	
	Heat island reduction, non-roof	2	
	Heat island reduction, roof	2	
	Outdoor light pollution reduction	1	
	Universal design	1	
	Basic facilities for construction workforce	1	
	Green building guidelines	1	
Water Conservation (WC)	Rainwater harvesting, roof and non-roof	Required	19
	Water-efficient plumbing fixtures	Required	
	Landscape design	2	
	Management of irrigation systems	1	
	Rainwater harvesting, roof & non-roof	4	
	Water-efficient plumbing fixtures	5	
	Wastewater treatment and reuse	5	
	Water metering	2	

(Continued)

Table 7.20 (Continued)

Assesment category	Assessment criteria	Credits available	Total credits
Energy Efficiency (EE)	Ozone-depleting substances	Required	28
	Minimum energy efficiency	Required	
	Commissioning plan for building equipment and systems	Required	
	Eco-friendly refrigerants	1	
	Enhanced energy efficiency	15	
	On-site renewable energy	6	
	Off-site renewable energy	2	
	Commissioning, post-installation of equipment and systems	2	
	Energy metering and management	2	
Building Materials and Resources (BMR)	Segregation of waste, post-occupancy	Required	16
	Sustainable building materials	8	
	Organic waste management, post-occupancy	2	
	Handling of waste materials during construction	1	
	Use of certified green building materials, products, and equipment	5	
Indoor Environmental Quality (IEQ)	Minimum fresh air ventilation	Required	12
	Tobacco smoke control	Required	
	CO2 monitoring	1	
	Daylighting	2	
	Outdoor views	1	
	Minimise indoor and outdoor pollutants	1	
	Low-emitting materials	3	
	Occupant well-being facilities	1	
	Indoor air quality testing, after construction and before occupancy	2	
	Indoor air quality management, during construction	1	
Innovation and Development (ID)	Innovation in design process	4	7
	Optimisation in structural design	1	
	Wastewater reuse, during construction	1	
	IGBC accredited professional	1	

Source: IGBC (2014).

the following eight categories, namely Management (MN), Sustainable Sites (SS), Energy and Atmosphere (EA), Materials and Resources (MR), Indoor Environmental Quality (EQ), Innovation and Design Process (ID), and Social and Cultural Awareness (SC). Based on the total credit points achieved after evaluation, certification can be awarded. The certification benchmarks for the GREEN[SL] rating system rate a project 'Certified Rated Green Buildings' with a percentage score of 40 to 49, 'Silver Rated Green Buildings' with a percentage score of 50 to 59, 'Gold Rated Green Buildings' with a percentage score of 60 to 69, and 'Platinum Rated Green Buildings' with a percentage score of 70 points and above.

Israeli Green Building Standard SI 5281

The SI 5281: Sustainable Building (also known as SI 5281: Buildings of Lesser Environmental Harm), SI 1045: Thermal Insulation of Buildings, and SI 5282: Energy Rating of Buildings are the three GB standards in Israel. The SI 5281: Sustainable Building rating system was developed in 2005 and extensively upgraded in 2011 by the Israeli Green Building Council (ILGBC), the Ministry of Building and Housing, the Ministry of Interior, the Standards Institute of Israel, and the Ministry of Environmental Protection (MEP). According to Shaviv and Pushkar (2014), the revised standard which consists of nine evaluation categories addresses different building types such as public, residential, retail, office, healthcare, education, and hotels. The categories addressed in the standard include Energy, Land, Water, Waste, Health and Well Being, Materials, Environmental Management, Transportation, and Innovation (Standards Institution of Israel, 2011; MEP State of Israel, 2012).

Home Performance Index (HPI)

The Home Performance Index by the Irish Green Building Council (IGBC) is the first national voluntary rating tool aimed at achieving quality and sustainable residential development in Ireland (IGBC, 2020). Based on the overall credit points attained across all evaluation categories, certification is awarded resulting in a Good, Better, or Best Practice level. However, two levels of certification are obtainable in the HPI rating system, namely 'Certified' (for scores ≥ 45%) and 'Gold' (for scores ≥ 65%) based on the total scores achieved and mandatory performance scores from certain indicators (IGBC, 2019). The HPI rating system evaluates the sustainability performance of residential buildings across five categories Environment, Health and Well-being, Economic, Quality Assurance, and Sustainable Location. The Environment category and its indicators assess the ecological footprint of the building project, the Health and Wellbeing category and its indicators assess users' daily dwelling experience and the impact on their overall wellbeing, the Economic category and its indicators assess the users' long-term value stability and running cost of the building project, the Quality Assurance category and its indicators assess the design and construction process of the building project while the Sustainable Location category and its indicators evaluate the effective relationship between the building project and access to existing transport infrastructure and amenities (IGBC, 2019).

Table 7.21 ARZ Buildings rating system evaluation categories

Module	Categories/Item	Possible credit points
M1	Energy Performance	6
M2	Thermal Energy	38
M3	Electrical Energy	33
M4	Building Envelope	36
M5	Materials	8
M6	Indoor Air Quality	9
M7	Operation and Management	11
M8	Water Conservation	9
M9	Bonus	16

Source: LGBC and IFC (2020).

ARZ Building Rating System

The ARZ building rating system was developed by the Lebanon Green Building Council (LGBC) in partnership with the International Finance Corporation (IFC) as the first GB rating tool for Lebanon. The intention was to create and promote a sustainability consciousness in the design, planning, construction, and operation of new and existing buildings to achieve a sustainable built environment in Lebanon (Ibrahim, 2017). According to the LGBC (2020), the development of the ARZ rating system is aimed at evaluating the sustainability performance of commercial buildings in Lebanon looking at the health, workplace comfort, and efficient resource consumption of the buildings. There are nine evaluation categories/modules (eight core modules and one bonus module) within the ARZ rating system from which a building can obtain 150 credit points (from the eight core modules) and another 16 points (from the bonus item). As shown in Table 7.21, the nine evaluation categories/modules are Energy Performance, Thermal Energy, Electrical Energy, Building Envelope, Materials, Indoor Air Quality, Operation and Management, Water Conservation, and Bonus. With a minimum required score of 135 points, 'Gold' certification is awarded, 'Silver' certification is awarded for a minimum required score of 120 points, 'Bronze' certification is awarded for a minimum required score of 100 points, 'Certified' certification is awarded for a minimum required score of 80 points, and 'Non-Certified' is awarded for a minimum required score of less than 80 points.

Building Sustainability Assessment Tools Development Research

There is a growing need to adopt and implement sustainable measures to curb the adverse environmental impacts of the built environment leading to the development of GBRTs or SATs. Despite the high number of these tools in existence and used globally, researchers are working tirelessly to develop new tools that address various building types and projects, and the specific sustainability demands of each country/region. For example, the study of Markelj et al. (2014) presents the development of a simplified method for evaluating building sustainability (SMEBS) in the early design phase for architects. The study of Olawumi et al. (2020) developed a building sustainability assessment method (BSAM) for developing countries in sub-Saharan Africa. A novel framework was developed by Alawneh et al. (2019) for integrating the UN SDGs into sustainable non-residential building assessment and management in Jordan. The study of Cetiner and Edis (2014) presented an environmental and economic sustainability assessment method for the retrofitting of residential buildings in Turkey. A Building information modelling (BIM)-based Kazakhstan building sustainability assessment framework (KBSAF) was developed from the study by Akhanova et al. (2020). The study of Akadiri et al. (2012) was aimed at developing a conceptual framework for implementing sustainability in the UK building sector while the study of Malek and Grierson (2016) was aimed at developing a building sustainability assessment method (BSAM) for Iran. The study of Saraiva et al. (2019) adapted the SBTool to develop a sustainability assessment of high school buildings (SAHSB) in Portugal while the study of Castro et al. (2017) developed a healthcare building sustainability assessment method (HBSAtool-PT) for Portugal.

The SMEBS developed in the early design phase for architects by Markelj et al. (2014) consists of 10 categories and 33 criteria that address the triple dimensions of SD. The evaluation categories are Pollution and Waste, Energy, Water Use, Materials, Sustainable Land Use, Well-being, Functionality, Technical Characteristics, Costs, and Property Value. The BSAM for developing countries in sub-Saharan Africa by Olawumi et al. (2020) evaluates the sustainability performance of building projects using 8 indicators/categories and 32 criteria. The sustainability indicators assessed are Sustainable Construction Practices, Site and Ecology, Energy, Water, Material and Waste, Transportation, Indoor Environmental Quality, and Building Management. The framework developed by Alawneh et al. (2019) for integrating the UN SDGs into sustainable non-residential building assessment and management in Jordan consists of 12 assessment categories and 75 assessment indicators. The assessment categories are Energy, Water, Indoor Environmental Quality, Material, Sustainable Site, Transport, Management, Waste, Pollution, Economic, Quality of Services, and Social and Cultural Value.

The BIM-based KBSAF developed by Akhanova et al. (2020) consists of 9 assessment categories and 45 assessment indicators. The assessment categories are Construction Site Selection and Infrastructure, Building Architectural and Planning Solutions Quality, Indoor Environmental Quality and Comfort, Water Efficiency, Energy Efficiency, GB Materials, Waste, Economy, and Management The conceptual framework for implementing sustainability in the UK building sector developed by Akadiri et al. (2012) consists of 9 assessment categories and 44 assessment indicators. The assessment categories are Energy Conservation, Material Conservation, Water Conservation, Land Conservation, Initial Cost, Cost in Use, Recovery Cost, Protecting Human Health and Comfort, and Protecting Physical Resources. The SAHSB developed for Portugal by Saraiva et al. (2019) consists of 23 indicators/criteria that are aggregated into 11 assessment categories. The assessment categories are Climate Change and Outdoor Air Quality, Biodiversity/Land Use, Energy, Materials, Solid Residues and Resources Management, Water,

User Health and Comfort, Accessibility, Security and Safety, Education for Sustainability Awareness, Sustainability of the Area, and Life Cycle Costs. Finally, the HBSAtool-PT developed for Portugal as adapted from the SBTool by Castro et al. (2017) consists of 22 assessment categories and 44 assessment indicators. The assessment categories which are classified into sustainability areas of environmental, sociocultural and functional, economy, technical, and site are Environmental Life Cycle Impact Assessment, Energy, Soil Use and Biodiversity, Materials and Solid Waste, Water, User's Health and Comfort, Controllability by the User, Landscaping, Passive Design, Mobility Plan, Space Flexibility and Adaptability, Life Cycle Costs, Local Economy, Environmental Management Systems, Technical Systems, Security, Durability, Awareness and Education for Sustainability, Skills in Sustainability, Local Community, Cultural Value, and Facilities.

Summary

This chapter extensively examines the various tools and systems for assessing GBs globally. The identification and critical examination of the constituting standardised benchmarks and checklists that made up these tools form the basis upon which the conceptual framework for this book was achieved. To formulate a tool that is truly sustainable, user-friendly, and cost-effective, these criteria are imperative to the effective performance of the proposed biomimicry sustainability assessment tool. Notable and widely used tools such as BREEAM, LEED, CASBEE, and Green Star among others are examined in this chapter. Finally, this chapter discusses the research development of a few building sustainability assessment tools identified in the literature. The next chapter will review the various green initiatives and building sustainability assessment tools that are operational in South Africa.

References

AbdelAzim, A. I., Ibrahim, A. M., & Aboul-Zahab, E. M. (2017). Development of an energy efficiency rating system for existing buildings using analytic hierarchy process – The case of Egypt. *Renewable and Sustainable Energy Reviews, 71*, 414–425.

Abrahams, G. (2017). Constructing definitions of sustainable development. *Smart and Sustainable Built Environment, 6*(1), 34–47.

Abu Dhabi Urban Planning Council (2010). *The pearl rating system for Estidama: Building rating system design & construction version 1.0*. Abu Dhabi, UAE: Abu Dhabi Urban Planning Council.

Adegbile, M. B. (2013). Assessment and adaptation of an appropriate green building rating system for Nigeria. *Journal of Environment and Earth Science, 3*(1), 1–10.

Akadiri, P. O., Chinyio, E. A., & Olomolaiye, P. O. (2012). Design of a sustainable building: A conceptual framework for implementing sustainability in the building sector. *Buildings, 2*(2), 126–152.

Akhanova, G., Nadeem, A., Kim, J. R., & Azhar, S. (2020). A multi-criteria decision-making framework for building sustainability assessment in Kazakhstan. *Sustainable Cities and Society, 52*, 1–11.

Alawneh, R., Ghazali, F., Ali, H., & Sadullah, A. F. (2019). A novel framework for integrating United Nations sustainable development goals into sustainable non-residential building assessment and management in Jordan. *Sustainable Cities and Society, 49*, 1–20.

Alobaidi, K. A., Abdul Rahim, A. B., Mohammed, A., & Baqutayan, S. (2015). Sustainability achievement and Estidama green building regulations in Abu Dhabi vision 2030. *Mediterranean Journal of Social Sciences, 6*(4), 509–518.

Assaf, S., & Nour, M. (2015). Potential of energy and water efficiency improvement in Abu Dhabi's building sector – Analysis of Estidama Pearl rating system. *Renewable Energy, 82*, 100–107.

Awadh, O. (2017). Sustainability and green building rating systems: LEED, BREEAM, GSAS and Estidama critical analysis. *Journal of Building Engineering, 11*, 25–29.

Bahaudin, A. Y., Elias, E. M., & Saifudin, A. M. (2014). A comparison of the green building's criteria. *E3S Web of Conferences, 3*, 1–10.

Bahaudin, A. Y., Mohamed Elias, E., Nadarajan, S., Romli, R., Zainudin, N., & Mohd Saifudin, A. (2012). *An overview of the green building's criteria: Non-residential new construction.* In: *3rd International Conference on Technology and Operations Management (ICTOM2012),* 4th–6th July 2012, Bandung, West Java, Indonesia.

Bansal, S., Srijit Biswas, F., & Singh, S. K. (2015). Approach of fuzzy logic for evaluation of green building rating system. *International Journal of Innovative Research in Advanced Engineering, 2*(3), 35–39.

BEAM Society Limited. (2012). *About BEAM plus.* Retrieved 24 March 2020 from https://www.beamsociety.org.hk/en_about_us_0.php

Building and Construction Authority (2017). *Green mark for residential buildings: Technical guide and requirements.* Singapore: Green Mark Department, Building and Construction Authority.

Building Research Establishment Ltd. (2017). *BREEAM International New Construction 2016.* Retrieved 28 April 2019 from https://www.breeam.com/BREEAMInt2016SchemeDocument/#02_scope_newcon/scope.htm%3FTocPath%3D2.0%2520Scope%2520of%2520the%2520BREEAM%2520International%2520New%2520Construction%25202016%2520scheme%2520version%7C_____0

Building Research Establishment. (2018). *BREEAM.* Retrieved April 03 2018 from https://www.bre.co.uk/page.jsp?id=829

Castro, M. F., Mateus, R., & Bragança, L. (2017). Development of a healthcare building sustainability assessment method – Proposed structure and system of weights for the Portuguese context. *Journal of Cleaner Production, 148*, 555–570.

Cetiner, I., & Edis, E. (2014). An environmental and economic sustainability assessment method for the retrofitting of residential buildings. *Energy & Buildings, 74*, 132–140.

Chen, X. (2017). *Developing a new passive design approach for green building assessment scheme with integrated statistical analysis and optimization* (Doctoral dissertation). The Hong Kong Polytechnic University, Hong Kong.

Chen, X., Yang, H., & Lu, L. (2015). A comprehensive review on passive design approaches in green building rating tools. *Renewable and Sustainable Energy Reviews, 50*, 1425–1436.

Cheng, J. C., & Venkataraman, V. (2012). Collaborative system for HK-BEAM green building certification. In Y. Luo (Ed.), *Cooperative design, visualization and engineering* (pp. 211–218). Berlin: Springer.

Construction Specifications Institute (2013). *The CSI sustainable design and construction practice guide.* Somerset: John Wiley & Sons.

Deutsche Gesellschaft für Nachhaltiges Bauen. (2020). *Scheme overview*. Retrieved 22 March 2020 from https://www.dgnb-system.de/en/schemes/scheme-overview/

Ding, Z., Fan, Z., Tam, V. W. Y., Bian, Y., Li, S., Illankoon, I. M., Chethana S, & Moon, S. (2018). Green building evaluation system implementation. *Building and Environment, 133*, 32–40.

Doan, D. T., Ghaffarianhoseini, A., Naismith, N., Zhang, T., Ghaffarianhoseini, A., & Tookey, J. (2017). A critical comparison of green building rating systems. *Building and Environment, 123*, 243–260.

Eberl, S. (2010). DGNB vs. LEED: A comparative analysis. *Conference on Central Europe Towards Sustainable Building*, 1–5.

Energy Smart Communities Initiative. (2020). *Building for ecologically responsive design excellence (BERDE)*. Retrieved 24 March 2020 from https://www.esci-ksp.org/archives/project/building-for-ecologically-responsive-design-excellence-berde

Gandhi, S., & Jupp, J. (2014). BIM and Australian Green Star building certification. *Computing in Civil and Building Engineering, 1*, 275–282.

German Sustainable Building Council. (2020). *DGNB system*. Retrieved 23 February 2020, from https://www.dgnb.de/en/index.php

Gibberd, J. (2003). *Building systems to support sustainable development in developing countries*. Pretoria, South Africa: CSIR Building and Construction Technology.

Goodhew, S. (2016). *Sustainable construction processes: A resource text*. UK: John Wiley & Sons.

Government of Singapore. (2019). *BCA green mark scheme*. Retrieved 24 March 2020 from https://www.mnd.gov.sg/our-work/greening-our-home/bca-green-mark

Green Building Council Indonesia. (2020). *Greenship*. Retrieved 21 January 2020 from http://www.gbcindonesia.org/greenship

Green Building Council of Australia (2017). *Green star design & as built v1.2: Submission guidelines*. Australia: Green Building Council of Australia.

Green Building Council of Australia. (2020). *Green star: The what and why of certification*. Retrieved 24 March 2020 from https://new.gbca.org.au/green-star/

Green Building Council of Sri Lanka. (2020). *GREENSL® rating system for built environment*. Retrieved 24 March 2020 from https://www.srilankagbc.org/site/rating.php

Green Rating for Integrated Habitat Assessment Council, & The Energy and Resources Institute (2019). *GRIHA v. 2019 abridged manual*. India: TERI Press.

Gusta, S. (2016). Sustainable construction in Latvia – Opportunities and challenges. *15th International Scientific Conference*, 1291–1299.

Hui, E. C. M., Tse, C., & Yu, K. (2017). The effect of BEAM plus certification on property price in Hong Kong. *International Journal of Strategic Property Management, 21*(4), 384–400.

Ibrahim, I. A. S. (2017). Green architecture challenges in the Middle East within different rating systems. *Energy Procedia, 115*, 344–352.

Indian Green Building Council (2014). *IGBC green new buildings rating system*. India: Indian Green Building Council.

Indian Green Building Council. (2015). *IGBC rating systems*. Retrieved 22 January 2016 from https://igbc.in/igbc/redirectHtml.htm?redVal=showratingSysnosign

International Living Future Institute (2019). *Living building challenge 4.0: A visionary path to a regenerative future*. Seattle, USA: The International Living Future Institute.

International Living Future Institute. (2020). *Petals*. Retrieved 24 March 2020 from https://living-future.org/lbc/

Irish Green Building Council (2019). *Home performance index: Technical manual version 2.0*. Dublin: Irish Green Building Council.

Irish Green Building Council. (2020). *Home performance index.* Retrieved 24 March 2020 from https://www.igbc.ie/certification/hpi/

Japan Sustainable Building Consortium, & Institute for Building Environment and Energy Conservation. (2020). *Comprehensive assessment system for built environment efficiency.* Retrieved 23 February 2020 from http://www.ibec.or.jp/CASBEE/english/index.htm

Kibert, C. J. (2016). *Sustainable construction: Green building design and delivery* (4th ed.). Hoboken, NJ: John Wiley and Sons Inc.

Kim, H., & Park, W. (2018). A study of the energy efficiency management in green standard for energy and environmental design (g-SEED)-certified apartments in South Korea. *Sustainability, 10*(3402), 1–20.

Kubba, S. (2012). *Handbook of green building design and construction: LEED, BREEAM, and green globes.* USA: Butterworth-Heinemann.

Kubba, S. (2016). *LEED v4 practices, certification, and accreditation handbook* (2nd edition. ed.). US: Butterworth-Heinemann.

Lebanon Green Building Council. (2020). *ARZ building rating system.* Retrieved 24 January 2020 from https://lebanon-gbc.org/arz-building-rating-system/.

Lee, W. L., & Burnett, J. (2008). Benchmarking energy use assessment of HK-BEAM, BREEAM and LEED. *Building and Environment, 43*(11), 1882–1891.

Lee, J., & Shepley, M. (2019). The green standard for energy and environmental design (g-seed) for multi-family housing rating system in Korea: A review of evaluating practices and suggestions for improvement. *Journal of Green Building, 14*(2), 155–175.

Li, Y., Chen, X., Wang, X., Xu, Y., & Chen, P. (2017). A review of studies on green building assessment methods by comparative analysis. *Energy and Buildings, 146*, 152–159.

Mah, D. E. (2011). *Framework for rating the sustainability of the residential construction practice.* Canada: University of Alberta.

Malek, S., & Grierson, D. (2016). A contextual framework for the development of a building sustainability assessment method for Iran. *Open House International, 41*(2), 64–75.

Markelj, J., Kitek Kuzman, M., Grošelj, P., & Zbašnik-Senegačnik, M. (2014). A simplified method for evaluating building sustainability in the early design phase for architects. *Sustainability, 6*(12), 8775–8795.

Masara, V. M. (2019). The green certification adapted to Brazil: In search of efficiency and sustainability in buildings. *MOJ Civil Engineering, 5*(2), 69–72.

Ministry of Environmental Protection, State of Israel. (2012). *Green building standards in Israel.* Retrieved 24 March 2020 from http://www.sviva.gov.il/English/env_topics/GreenBuilding/Pages/GreenBuildingStandards.aspx

Mishra, A., & Kauškale, L. (2018). The analysis of the green building supporting organisations in the scandinavian countries and baltic States. *Baltic Journal of Real Estate Economics and Construction Management, 6*(1), 201–219.

Mitchell, L. M. (2010). Green star and NABERS: Learning from the Australian experience with green building rating tools. Energy Efficient Cities: Assessment Tools and Benchmarking Practices, 93–130. The International Bank for Reconstruction and Development/ The World Bank, Washington, DC, USA.

Mok, S., & Cho, D. (2014). Comparison criteria and certified scores between G-SEED and LEED certification systems in office building cases. *World SB14 Barcelona Conference*, 1–7.

Murakami, S., Kawakubo, S., Asami, Y., Ikaga, T., Yamaguchi, N., & Kaburagi, S. (2010). Concept and framework of CASBEE-city—an assessment tool of the built environment of cities. *Conference of Sustainable Building*, 1–5.

Newsham, G. R., Mancini, S., & Birt, B. J. (2009). Do LEED-certified buildings save energy? Yes, but. *Energy & Buildings, 41*(8), 897–905.

Ng, R., Shariff, F., Kristensen, P. E., & Zahry, M. (2007). Adapting SBTool as a sustainable building framework for Malaysia. *Proceeding of the International Conference on Sustainable Building*, 317–322.

Nguyen, B. K., & Altan, H. (2011). Comparative review of five sustainable rating systems. *Procedia Engineering, 21*, 376–386.

Olawumi, T. O., Chan, D. W. M., Chan, A. P. C., & Wong, J. K. W. (2020). Development of a building sustainability assessment method (BSAM) for developing countries in sub-Saharan Africa. *Journal of Cleaner Production, 263*, 1–37.

Olgyay, V., & Herdt, J. (2004). The application of ecosystems services criteria for green building assessment. *Solar Energy, 77*(4), 389–398.

Patenaude, M., & Plouffe, S. (2015). Consideration of the use phase in certification programs for residential green building. *Journal of Green Building, 10*(1), 150–168.

Philippine Green Building Council. (2015). *Building ecologically responsive design excellence.* Retrieved 12 December 2019 from http://philgbc.org/tag/building-ecologically-responsive-design-excellence/

Philippine Green Building Council. (2020). *About BERDE.* Retrieved 24 March 2020 from https://berdeonline.org/#about-berde

Qadir, G., Haddad, M., & Hamdan, D. (2019). Potential of energy efficiency for a traditional Emirati house by Estidama Pearl rating system. *Energy Procedia, 160*, 707–714.

Rastogi, A., Choi, J., Hong, T., & Lee, M. (2017). Impact of different LEED versions for green building certification and energy efficiency rating system: A multifamily midrise case study. *Applied Energy, 205*, 732–740.

Reeder, L. (2010). *Guide to green building rating systems: Understanding LEED, green globes, Energy Star, the national green building standard, and more.* Hoboken, New Jersey: John Wiley & Sons.

Sande, I. I., & Phadtare, N. S. (2015). Comparative study of LEED and GRIHA rating system. *Journal of Information, Knowledge and Research in Civil Engineering, 3*(2), 168–176.

Saraiva, T. S., Almeida, M., & Bragança, L. (2019). Adaptation of the SBTool for sustainability assessment of high school buildings in Portugal – SAHSBPT. *Applied Sciences, 9*(13), 1–15.

SAS International. (2020). *Estidama pearl building rating system.* Retrieved 14 January 2020 from https://sasintgroup.com/info-hub/environmental-accreditation-statements/estidama-pearl-building-rating-system/

Sasatani, D., Bowers, T., Ganguly, I., & Eastin, I. L. (2015). Adoption of CASBEE by Japanese house builders. *Journal of Green Building, 10*(1), 186–201.

Scofield, J. H. (2009). Do LEED-certified buildings save energy? Not really. *Energy & Buildings, 41*(12), 1386–1390.

Scofield, J. H. (2013). Efficacy of LEED certification in reducing energy consumption and greenhouse gas emission for large New York City office buildings. *Energy and Buildings, 67*, 517–524.

Sev, A. (2011). A comparative analysis of building environmental assessment tools and suggestions for regional adaptations. *Civil Engineering and Environmental Systems, 28*(3), 231–245.

Shamseldin, A. K. M. (2018). Including the building environmental efficiency in the environmental building rating systems. *Ain Shams Engineering Journal, 9*(4), 455–468.

Sharifi, A., Murayama, A., & Nagata, I. (2012). The potential of "CASBEE for urban development" for delivering sustainable communities: A case study from the "Koshigaya lake town" planning experience. *International Symposium of Urban Planning, 1*, 703–713.

Shaviv, E., & Pushkar, S. (2014). Green building standards – visualization of the building as layers according to lifetime expectancy. *Energy Procedia, 57*, 1696–1705.

Shukla, A., Singh, R., & Shukla, P. (2015). Achieving energy sustainability through green building approach. In *Energy sustainability through green energy* (pp. 147–162). India: Springer.

Shwe, T., Homma, R., Iki, K., & Ito, J. (2017). The potential of 'Comprehensive assessment system for built environment efficiency for cities' in developing country: Evidence of Myanmar. *International Journal of Architectural and Environmental Engineering, 11*(4), 523–531.

Siebers, R., Kleist, T., Lakenbrink, S., Bloech, H., den Hollander, J., & Kreißig, J. (Eds.). (2016). *Sustainability certification labels for buildings*. UK: Wiley Online Library.

Srivastava, A., Magar, R. B., & Shah, D. S. (2017). A review of IGBC rating system for new green buildings. *International Conference on Emanations in Modern Technology and Engineering (ICEMTE-2017), 5*(3), 14–17.

Standards Institution of Israel (2011). *SI 5281 – Sustainable buildings: Part 2 – Requirements for residential buildings*. Israel: Standards Institution of Israel.

Suzer, O. (2015). A comparative review of environmental concern prioritization: LEED vs other major certification systems. *Journal of Environmental Management, 154*, 266–283.

Tae, S., & Shin, S. (2009). Current work and future trends for sustainable buildings in South Korea. *Renewable and Sustainable Energy Reviews, 13*(8), 1910–1921.

Tambovceva, T., Geipele, I., & Geipele, S. (2012). Sustainable building in Latvia: Development and future challenges. *ISEE 2012 Conference–Ecological Economics and Rio 20*: Abstracts and Full Papers; 1–10.

Thiel, C., Campion, N., Landis, A., Jones, A., Schaefer, L., & Bilec, M. (2013). A materials life cycle assessment of a net-zero energy building. *Energies, 6*(2), 1125–1141.

Thilakaratne, R., & Lew, V. (2011). Is LEED leading Asia? An analysis of global adaptation and trends. *Procedia Engineering, 21*, 1136–1144.

Turk, S., Quintana, S. N. S. A., & Zhang, X. (2018). Life-cycle analysis as an indicator for impact assessment in sustainable building certification systems: The case of Swedish building market. *Energy Procedia, 153*, 414–419.

Uğur, L. O., & Leblebici, N. (2018). An examination of the LEED green building certification system in terms of construction costs. *Renewable and Sustainable Energy Reviews, 81*, 1476–1483.

United States Green Building Council. (2019). *Leadership in energy and environmental design*. Retrieved 09 August 2019 from https://new.usgbc.org/leed

Vij, A. (2010). National green building assessment tool in India. *Prague CESB10 – Central Europe towards Sustainable Building Conference*, 1–12.

Vyas, S., Ahmed, S., & Parashar, A. (2014). BEE (Bureau of Energy Efficiency) and green buildings. *International Journal of Research, 1*(3), 23–32.

Weerasinghe, U. G. D. (2012). *Development of a framework to assess sustainability of building projects*. Canada: University of Calgary.

Wong, S., & Abe, N. (2014). Stakeholders' perspectives of a building environmental assessment method: The case of CASBEE. *Building and Environment, 82*, 502–516.

Wu, Z., Li, H., Feng, Y., Luo, X., & Chen, Q. (2019). Developing a green building evaluation standard for interior decoration: A case study of China. *Building and Environment, 152*, 50–58.

Wu, P., Mao, C., Wang, J., Wang, X., & Song, Y. (2016). A decade review of the credits obtained by LEED v2.2 certified green building projects. *Building and Environment, 102*, 167–178.

Wu, M., & Yau, R. (2005). A comprehensive environmental performance assessment scheme for buildings in Hong Kong. *The World Sustainable Building Conference*, Tokyo, 1822–1829.

Xia, B., Zuo, J., Skitmore, M., Pullen, S., & Chen, Q. (2013). Green star points obtained by Australian building projects. *Journal of Architectural Engineering, 19*(4), 302–308.

Yeom, D., & Lee, K. (2015). Study on the improvement of Korean green building certification criteria focused on certification score and specialist survey analysis. *Journal of Asian Architecture and Building Engineering, 14*(1), 129–136.

Yun, Y., Cho, D., & Chae, C. (2018). Analysis of green building certification system for developing g-SEED. *Future Cities and Environment, 4*(1), 1–9.

Zarghami, E., Azemati, H., Fatourehchi, D., & Karamloo, M. (2018). Customizing well-known sustainability assessment tools for Iranian residential buildings using fuzzy analytic hierarchy process. *Building and Environment, 128*, 107–128.

Zhang, Y. E. (2015). *Research on green building assessment system in China inspired by LEED V4 and other foreign assessment system*. University of Florida.

Zimmermann, R. K., Skjelmose, O., Jensen, K. G., Jensen, K. K., & Birgisdottir, H. (2019). Categorizing building certification systems according to the definition of sustainable building. *IOP Conference Series: Materials Science and Engineering, 471*, 1–8.

8 Green Initiatives and Building Sustainability Assessment in South Africa

Understanding the Built Environment Terrain in South Africa

The construction industry (CI) as a labour-intensive sector creates employment while delivering roads, buildings, factories, electricity and water supply infrastructures, and other amenities that enhance the socio-economic status of the citizens (James et al., 2012). As a nation plagued with poverty and a significant increase in youth unemployment resulting in a high crime rate, the South African Government is continuously developing and adopting ways of addressing these challenges. The government's recognition of the benefits associated with the CI has therefore resulted in heavy investment in infrastructural development which has a major direct linkage to the built environment. With the undeniable prevalence of an intimate relationship between unemployment and poverty (in rural and urban areas) throughout the country (Ross et al., 2010; Zizzamia, 2020), the potential of the CI to close the widened poverty and unemployment gap will be realised.

The CI globally has been identified as responsible for causing several negative environmental issues, mainly climate change and many more. The sector is perceived as risky and hazardous owing to its associated issues and challenges such as accidents and injuries, resource consumption, waste generation, pollution, and emissions, amongst numerous others. Though some of these issues/challenges have been in existence for a while, there is a paucity of information to inform that they are completely eradicated and non-existent (Windapo & Cattell, 2013). To address these impacts, the concept of sustainability or sustainable development (SD) is adopted and integrated into the CI. Since the United Nations (UN) conference of 1972 was held in Stockholm, the issue of SD became a global concern (Windapo, 2014). By adopting sustainable construction (SC) or green building (GB), the adverse environmental impacts of the CI will be minimised, resulting in the improved socio-economic status of local communities and the general populace (Gunnel, 2009). To this end, South Africa, since its independence in 1994, has committed to SD mechanisms as evidenced in the legislation and policies of the country (Ross et al., 2010).

DOI: 10.1201/9781003415961-11

Sustainability Drives, Policies, and Initiatives in South Africa

There are several initiatives, action plans, institutional policies, educational movements and collaborations with the private sector and industry giants developed and implemented by the South African government (Gupta & Laubscher, 2018) towards achieving SD in South Africa. Most of these policies and initiatives are propelled by the government while others are spearheaded by non-governmental organisations (NGOs) and agencies. This section, therefore, presents an overview of the efforts geared towards achieving South Africa's vision for a sustainable society.

South Africa National Framework for Sustainable Development

The UN Conference on Environment and Development (Rio de Janeiro Earth Summit) of 1992 and the Johannesburg World Summit on SD held in 2002 provided an avenue and foundation to understand and begin the implementation of sustainability practices in South Africa (Department of Forestry, Fisheries and the Environment, 2019). As inferred from these conferences, poverty is identified as a major challenge facing Africa because most countries on the continent are marginalised and do not benefit from the opportunities that come with globalisation (United Nations, 2004). Based on this, the Cabinet in 2008 assented to the creation of the South Africa National Framework for Sustainable Development (NFSD), heralding a new era of thinking targeted at promoting the sustainability of South Africa's economic, social, and natural resources.

The NFSD is aimed at enunciating 'South Africa's national vision for SD and indicating strategic interventions to re-orientate South Africa's development path in a more sustainable direction' (United Nations, 2020). To achieve this aim, the following five strategic pathways are established:

a   Enhancing systems for integrated planning and implementation;
b   Sustaining our ecosystems and using natural resources efficiently;
c   Investing in sustainable infrastructure to achieve economic development;
d   Creating sustainable human settlements; and
e   Responding appropriately to emerging human development, economic, and environmental challenges (Department of Environmental Affairs and Tourism, 2008).

National Strategy for Sustainable Development and Action Plan

To provide a step-by-step path to the realisation of the NFSD, the Cabinet in 2011 approved the first National Strategy for Sustainable Development and Action Plan (NSSD 1) which was meant for implementation from 2011 to 2014. This is in tandem with South Africa's vision for a sustainable society in which 'South Africa aspires to be a sustainable, economically prosperous and self-reliant nation that safeguards its democracy by meeting the fundamental human needs of its people, by managing its limited ecological resources responsibly for current and future generations, and by advancing efficient and effective integrated planning and governance through national, regional and global collaboration' (Department of Environmental Affairs and Tourism, 2008).

Five strategic priorities and 20 headline indicators as an associated action plan within the context of SD are developed and contained in the NSSD 1. The reformulated NSSD 1 strategic priorities are: Enhancing systems for integrated planning and implementation (Priority 1), Sustaining our ecosystems and using natural resources efficiently (Priority 2), Towards a green economy (Priority 3), Building sustainable communities (Priority 4), and Responding effectively to climate change (Priority 5). The strategic priorities and their corresponding headline indicators are presented as follows:

Strategic Priority 1: Enhancing Systems for Integrated Planning and Implementation.

- Establish an effective National Committee on Sustainable Development (NCSD).
- Increase the number of government entities and private sector companies that report against sustainability indicators (King III sustainability reporting, Carbon Disclosure Project, and Water Disclosure Project).
- Increase the number and begin measuring community-based capacity-building projects.

Strategic Priority 2: Sustaining Our Ecosystems and Using Natural Resources Efficiently.

- Curtail water losses at water distribution systems to an average percentage reduction (from 30 to 15% by 2014).
- Reduction of demand as determined in the reconciliation strategies for seven large water supply systems by 15% (assessment of water requirements and water monitoring systems implemented by 2014).
- Increase the number of Blue Flag beaches (to above 29 beaches).
- Rehabilitate land affected by degradation (3.2 million ha by 2014).
- Increase the percentage of coastline with partial protection (from 12% to 14% by 2014).
- Increase the percentage of land mass protected (formal and informal) from 6.1% to 9% by 2014.

Strategic Priority 3: Towards a Green Economy

- Ensure progress on the implementation of the nine green economy programmes with an impact on social (jobs), economic (industry development), and environmental (ecosystem) benefits by 2014.
- Increase percentage (or amount) of financial resources ringfenced/streamlined and spent for green economy programmes (2010/11 amount – Industrial Development Corporation: R11.7 billion, Development Bank of South Africa: R25 billion, Private: >R100 billion, National Treasury: R800 million).

- Increase the number of patents, prototypes, and technology demonstrators added to the intellectual property (IP) portfolio annually from funded or co-funded research programmes (five additions to the IP portfolio – patents, patent applications, licenses, and trademarks – by March 2014).
- Share of GDP of the Environmental Goods and Services (EGS) Sector (3% of GDP by 2014).

Strategic Priority 4: Building Sustainable Communities.

- Increase the percentage of households with access to water (92 to 100%), sanitation (69 to 100%), refuse removal (64 to 75%), and electricity (81 to 92%).
- Upgrade 400 000 households in well-located informal settlements with access to basic services and secure tenure (approximately 2700 informal settlements are in good locations, i.e., located close to metropolitan areas and basic services, have high densities and, in 2008, housed approximately 1.2 million households).
- Increase the South African Human Development Index (HDI) [2010 HDI: 0.597].
- Gini coefficient (reduce income inequality) [2008: 0.66].

Strategic Priority 5: Responding Effectively to Climate Change.

- Reduce greenhouse gas emissions (metric ton CO_2 equivalent) [34% reduction below a business-as-usual baseline by 2020 and 42% by 2025].
- Increase the percentage of power generation that is renewable (10,000 GWh by 2014).
- Improve climate change adaptation plans developed with 12 sectors by 2012 (Biodiversity, Forestry, Water, Coastal Management, Agriculture, Health, Tourism, Land and Rural Development, Local Government, Fisheries, Human Settlements, Business/Insurance).

National Development Plan 2030

South Africa's National Development Plan (NDP) 2030 was unveiled in 2013 by the Minister of the Presidency: National Planning Commission at a media briefing on its implementation. As summarised by the Green Growth Knowledge Platform (2020), the NDP was established with the following six main objectives:

a Uniting all South Africans around a common programme to achieve prosperity and equity;
b Promoting active citizenry to strengthen development, democracy, and accountability;
c Bringing about faster economic growth, higher investment, and greater labour absorption;
d Focusing on key capabilities of people and the state;

e Building a capable and developmental state; and
f Encouraging strong leadership throughout society to work together to solve problems.

The broad aim of the aforementioned objectives of the NDP is to eliminate poverty and reduce inequality by the year 2030 which has seen a new call to action and a brand identity launched in 2017 (Government of South Africa, 2020). At the centre of the NDP is the aim to ensure a decent standard of living is achieved for all South Africans by 2030. As listed by the National Planning Commission (NPC, 2020), a decent standard of living should incorporate the following vital elements, namely housing, water, electricity, and sanitation; safe and reliable public transport; quality education and skills development; safety and security; quality healthcare; social protection; employment; recreation and leisure; a clean environment; and adequate nutrition. Figure 8.1 provides a summary of the elements of a decent standard of living. To ensure the successful implementation of the 2030 NDP, 13 major areas are identified with corresponding objectives and actions highlighted against them. The areas include economy and employment; economic infrastructure; environmental sustainability and resilience; inclusive rural economy; South Africa in the region and the world; transforming human settlements; improving education, training, and innovation; healthcare for all; social protection; building safer communities; building a capable and developmental state; fighting corruption; and nation building and social cohesion. Table 8.1 presents a summary of the objectives of the 2030 NDP major areas.

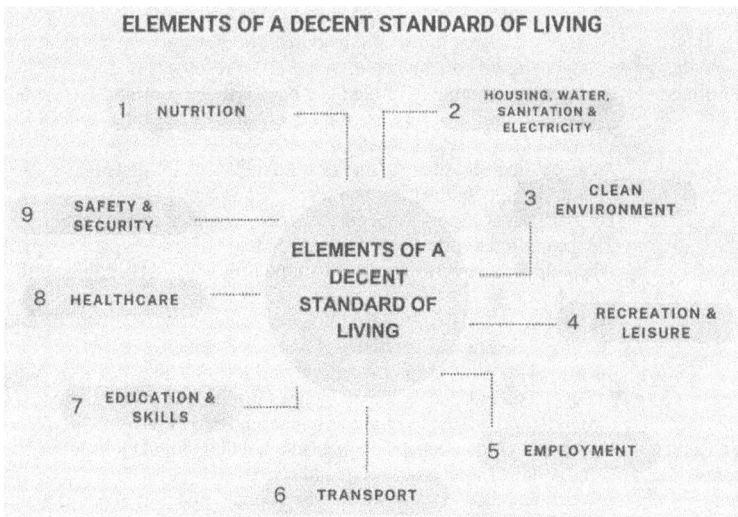

Figure 8.1 Summary of the elements of a decent standard of living

Table 8.1 Summarised areas and objectives of 2030 NDP

Strategic priorities	Objectives
Economy and Employment	The unemployment rate should fall from 24.9% in June 2012 to 14% by 2020 and to 6% by 2030. This requires an additional 11 million jobs. Total employment should rise from 13 million to 24 million
	The proportion of adults working should increase from 41% to 61%
	The proportion of adults in rural areas working should rise from 29% to 40%
	The labour force participation rate should rise from 54% to 65%
	Gross domestic product (GDP) should increase by 2.7 times in real terms, requiring average annual GDP growth of 5.4% over the period. GDP per capita should increase from about from about R50 000 per person in 2010 to R110 000 per person in 2030 at constant prices
	The proportion of national income earned by the bottom 40% should rise from about 6% today to 10% in 2030
	Broaden ownership of assets to historically disadvantaged groups
Economic Infrastructure	The proportion of people with access to the electricity grid should rise to at least 90% by 2030, with non-grid options available for the rest
	The country would need an additional 29 000MW of electricity by 2030. About 10 900MW of existing capacity is to be retired, implying new build of more than 40 000MW
	At least 20 000MW of this capacity should come from renewable sources
	Ensure that all people have access to clean, potable water and that there is enough water for agriculture and industry, recognising the trade-offs in the use of water
	Reduce water demand in urban areas to 15% below the business-as-usual scenario by 2030
	The proportion of people who use public transport for regular commutes will expand significantly. By 2030, public transport will be user-friendly, less environmentally damaging, cheaper and integrated or seamless
	Durban port capacity should increase from 3 million containers a year to 20 million by 2040
	Competitively priced and widely available broadband
Environmental Sustainability and Resilience	A set of indicators for natural resources, accompanied by publication of annual reports on the health of identified resources to inform policy
	A target for the amount of land and oceans under protection (presently about 7.9 million hectares of land, 848kms of coastline, and 4 172 square kilometres of ocean are protected)
	Achieve the peak, plateau, and decline trajectory for greenhouse gas emissions, with the peak being reached around 2025
	By 2030, an economy-wide carbon price should be entrenched
	Zero emission building standards by 2030
	Absolute reductions in the total volume of waste disposed to landfill each year
	At least 20 000MW of renewable energy should be contracted by 2030
	Improved disaster preparedness for extreme climate events
	Increased investment in new agricultural technologies, research, and the development of adaptation strategies for the protection of rural livelihoods and expansion of commercial agriculture
Inclusive Rural Economy	An additional 643 000 direct jobs and 326 000 indirect jobs in the agriculture, agro-processing, and related sectors by 2030
	Maintain a positive trade balance for primary and processed agricultural products

(Continued)

Table 8.1 (Continued)

Strategic priorities	Objectives
South Africa in the Region and the World	Intra-regional trade in Southern Africa should increase from 7% of trade to 25% of trade by 2030
	South Africa's trade with regional neighbours should increase from 15% of our trade to 30%
Transforming Human Settlements	Strong and efficient spatial planning system, well integrated across the spheres of government
	Upgrade all informal settlements on suitable, well-located land by 2030
	More people living closer to their places of work
	Better quality public transport
	More jobs in or close to dense, urban townships
Improving Education, Training and Innovation	Make early childhood development a top priority among the measures to improve the quality of education and long-term prospects of future generations. Dedicated resources should be channelled towards ensuring that all children are well cared for from an early age and receive appropriate emotional, cognitive, and physical development stimulation
	All children should have at least 2 years of pre-school education
	About 90% of learners in grades 3, 6, and 9 must achieve 50% or more in the annual national assessments in literacy, maths, and science
	Between 80% and 90% of learners should complete 12 years of schooling and/or vocational education with at least 80% successfully passing the exit exams
	Eradicate infrastructure backlogs and ensure that all schools meet the minimum standards by 2016
	Expand the college system with a focus on improving quality. Better quality will build confidence in the college sector and attract more learners. The recommended participation rate of 25% would accommodate about 1.25 million enrolments
	Provide 1 million learning opportunities through Community Education and Training Centres
	Improve the throughput rate to 80% by 2030
	Produce 30 000 artisans per year
	Increase enrolment at universities by at least 70% by 2030 so that enrolments increase to about 1.62 million from 950 000 in 2010
	Increase the number of students eligible to study towards maths and science-based degrees to 450 000 by 2030
	Produce more than 100 doctoral graduates per million per year by 2030. That implies an increase from 1420 in 2010 to well over 5 000 a year
	Expand science, technology, and innovation outputs by increasing research and development spending by government and through encouraging industry to do so
Health Care for All	Increase average male and female life expectancy at birth to 70 years
	Progressively improve TB prevention and cure
	Reduce maternal, infant, and child mortality
	Significantly reduce prevalence of noncommunicable chronic diseases
	Reduce injury, accidents, and violence by 50% from 2010 levels
	Deploy primary healthcare teams provide care to families and communities
	Everyone must have access to an equal standard of care, regardless of their income
	Fill posts with skilled, committed, and competent individuals

(Continued)

Table 8.1 (Continued)

Strategic priorities	Objectives
Social Protection	Ensure progressively and through multiple avenues that no one lives below a defined minimum social floor
	All children should enjoy services and benefits aimed at facilitating access to nutrition, health care, education, social care, and safety
	Address the skills deficit in the social welfare sector
	Provide income support to the unemployed through various active labour market initiatives such as public works programmes, training and skills development, and other labour market-related incentives
	All working individuals should make adequate provision for retirement through mandated savings. The state should provide measures to make pensions safe and sustainable
	Social protection systems must respond to the growth of temporary and part-time contracts, and the increasing importance of self-employment and establish mechanisms to cover the risks associated with such
	Create an effective social welfare system that delivers better results for vulnerable groups, with the state playing a larger role compared to now. Civil society should complement government initiatives
Building Safer Communities	In 2030 people living in South Africa feel safe and have no fear of crime. They feel safe at home, at school, and at work, and they enjoy an active community life free of fear. Women can walk freely in the street and the children can play safely outside. The police service is a well-resourced professional institution staffed by highly skilled officers who value their works, serve the community, safeguard lives and property without discrimination, protect the peaceful against violence, and respect the rights of all to equality and justice
Building a Capable and Developmental State	A state that can play a developmental and transformative role
	A public service immersed in the development agenda but insulated from undue political interference
	Staff at all levels have the authority, experience, competence, and support they need to do their jobs
	Relations between national, provincial, and local government are improved through a more proactive approach to managing the intergovernmental system
	Clear governance structures and stable leadership enable state-owned enterprises (SOEs) to achieve their developmental potential
Fighting Corruption	A corruption-free society, a high adherence to ethics throughout society, and a government that is accountable to its people
National Building and Social Cohesion	Our vision is a society where opportunity is not determined by race or birthright; where citizens accept that they have both rights and Responsibilities. Most critically, we seek a united, prosperous, non-racial, non-sexist, and democratic South Africa

Source: NPC (2020).

Department of Environment, Forestry, and Fisheries Projects and Programmes

As a major arm of government directly linked to sustainability issues, the department is saddled with the responsibility of initiating efforts, policies, and programmes aimed at driving South Africa's SD agenda. To this effect, several projects and programmes have been developed with the sole target of ensuring the country attains holistic sustainability. Examples of these projects and programmes include Working for Forests, Working for Ecosystems, Working for the Coast, Working for Water, Working for Land, Working for Wetlands, Working for Fire, Working on Waste, Youth Environmental Services (YES) Programme, Youth Jobs in Waste (YJW) Programme, Eco Furniture, Groen Sebenza Jobs Fund partnership project, Bioprospecting Economy, Biodiversity Economy, Access, and Benefit-Sharing (ABS) clearing house of the Republic of South Africa, People and Parks Programme, Kids in Parks Programme, Rhino Dialogues South Africa, Man and the Biosphere (MAB) Reserves Programme, Intergovernmental Science-Policy Platform on Biodiversity and Ecosystem Services (IPBES), Transfrontier Conservation Areas, Non-Motorised Transport (NMT)-South Africa, Strategic Environmental Assessment (SEA) for the Square Kilometre Array (SKA) Phase 1 South Africa, Green Cars, Green Fund, Green Economy for Sustainable Development, Donor funded projects, Greenest Municipality Competition (GMC), Green Passport, Greening and Open Space Management, Biomass Energy, EMI School Art competition, Operation Phakisa, Effective Environmental Improvement Interventions (2E2I), and Environment sector gender strategy. A synopsis of some of these policies and programmes is presented as documented by the Department of Forestry, Fisheries, and the Environment (2019) in a bid to provide a better comprehension of what they entail.

Working for Forest Programme

The idea of this programme is centred around rehabilitating the government's deteriorating remaining plantations and establishing an agency to manage and rehabilitate them. Under this programme, the government planned on having a new 100,000-ha afforestation in the KwaZulu-Natal and Eastern Cape provinces to address critical shortages in forest products (especially saw timber) and improve livelihood options for deep rural communities. The forest resources targeted are fuel woods, building materials, pulp and paper, poles, non-fibre products, and bio-renewable energy such as charcoal and biofuels. The vision of the programme is for a 'sustainable development and management of the newly initiated afforestation, transformed invading alien plant stands and degraded STATE forests into utilisable resources for the use of neighbouring communities in a manner that will maximize the socio-economic benefits, ensure the sustainable use and maintenance of natural areas and the minimize the risk of alien plant invasions' (Department of Forestry, Fisheries, and the Environment, 2019). The following objectives are highlighted in the implementation of the programme:

- Enhancing economic empowerment of deep rural communities through the development of value-added industries arising from plantation forests;
- Rehabilitating and sustainably managing natural resources and ecosystem function inside and surrounding plantation forests (e.g., fire regimes);
- Improving the productive potential of rural plantation forests and the creation of sustainable livelihoods;

- Creating exit opportunities for Working for Water contractors; and
- Optimising both the subsistence and commercial benefits from plantations forests and woodlots to society.

Working for Ecosystem Programme

According to the Department of Forestry, Fisheries and the Environment (2019), nearly 5.7 million hectares of untransformed land have been degraded in South Africa. It is based on this situation that a country-wide land restoration programme was developed to address the challenge and optimise the benefits of this land. The Working for Ecosystems programme is aimed at reversing environmental degradation, maintenance, and ecological restoration mechanisms. The following objectives are highlighted in the implementation of the programme:

- Improving watershed services through the restoration of watersheds (mountain catchment) services, riparian zones, and wetlands in collaboration with Working for Wetlands, Working for Water and Working on Fire programmes;
- Contributing to climate change mitigation through the sequestration of carbon in the form of re-vegetating denuded landscapes;
- Contributing to adaptation to the impacts of climate change and improving livelihood security by reducing the risk of natural disasters through the restoration of degraded habitats;
- Unlocking investments and operational resources for the improvement of the quantity and quality of ecosystem services as referred to in the first 3 points above; and
- Promoting pro-poor economic development in rural areas through the first 4 points above.

Working for the Coast Programme

A coalescence of effects of inland and marine coastal zone resource practices has characterised South Africa's coastlands. The identified issues that have occurred along the coastline are continuous sedimentation (direct killing of marine organisms, the declining attraction of beaches and coastal landscapes, and rise in local sea level and extensive coastal erosion, etc.), environmental pollution (pollution from agrochemical sources, pollution from domestic sources, industrial and oil pollution), direct destruction of coastal habitats (dredging activities, mining, land reclamation, and poor fishing methods), and urbanisation and the influx of tourists. To address these ills, the Working for the Coast programme was developed with the vision of a sustainable and healthy coastal environment that is equitably preserved and maintained for current and future generations. The following objectives are highlighted in the implementation of the programme:

- Equitable access to coastal public property; and
- Protection and conservation of coastal environment.

Working for Water Programme

Invasive alien species (microbes, animals and plants introduced into the country and then out-compete the indigenous ones) are causing billions of Rands of damage to South Africa's economy yearly. This issue is identified as the single major threat to the nation's biodiversity. To address this problem, the Working for Water (WfW) programme was launched in 1995 and has since created training and jobs for approximately 20,000 citizens from the most marginalised sectors of society yearly. The main objective of the programme is to reduce yearly by 22% the density of established terrestrial invasive alien plants via labour-intensive chemical, mechanical, and biological control. The methods of control adopted by field workers and scientists are highlighted as follows:

- Mechanical methods (felling, removing, or burning invading alien plants);
- Chemical methods (using environmentally safe herbicides);
- Biological control (using species-specific insects and diseases from the alien plant's country of origin which has seen 76 bio-control agents released against 40 weed species to date); and
- Integrated control (combinations of the above three approaches. Often an integrated approach is required to prevent enormous impacts).

Working for Land Programme

The Working for Land programme is a sustainable resource utilisation strategy based on cooperation and partnerships within communities. Necessitating the programme, unsustainable cutting of firewood, lack of grasslands protection, lack of stormwater management to reduce soil erosion and lack of educational programmes to inform communities on land management are identified as major challenges. The programme, however, aims to rehabilitate and restore degraded land and the composition structure of the environment. This is believed will reduce environmental risks, increase the productive capacity of the land, improve the sustainability of livelihoods, improve natural species diversity and catchment stability, and promote economic empowerment in rural areas. The following objectives are highlighted in the implementation of the programme:

- Promoting environmental education and awareness;
- Encouraging better land use practices;
- Mitigating loss of topsoil which will enhance the ecological integrity of the ecosystem;
- Curtailing of bush encroachment;
- Encouraging biodiversity conservation; and
- Restoring and rehabilitating degraded land.

Working for Wetlands Programme

Wetlands are identified as natural infrastructure and assets with the ability to freely provide a range of services, functions, and products. In South Africa, studies have shown that between 35% and 60% of wetlands are severely damaged or have already been lost. As a joint initiative of the Departments of Agriculture, Forestry, and Fisheries (DAFF), Water and Sanitation (DWS), and Environmental Affairs (DEA), the Water for Wetlands programme aims to propel the efficient utilisation of wetlands for poverty alleviation and employment creation. The following objectives are highlighted in the implementation of the programme:

- Communication, education, & public awareness;
- Knowledge sharing;
- Co-operative governance and partnerships;
- Skills and capacity development; and
- Wetland protection, wise use and rehabilitation.

Working on Fire Programme

Launched in September 2003 as part of the government's initiative to create jobs and alleviate poverty, the Working on Fire (WoF) programme focuses on the prevention and control of wildland fires to enhance the sustainability and protection of life, property and the environment through the implementation of integrated fire management (IFM) practices. The programme is aimed at enhancing the protection and sustainability of life, livelihoods, ecosystem services and natural processes through IFM. The following objectives are highlighted in the implementation of the programme:

- To reduce the risk of high biomass loads after clearing;
- To contribute to the reduction of densely invaded areas;
- To implement integrated fire management;
- To prevent and control wildfires; and
- To reduce the cost of clearing the invasive plants.

Working on Waste Programme

As one of the initiatives of the DEA implemented under the auspices of the Expanded Public Works Programme (EPWP), the Working on Waste programme is a proactive preventive measure that identifies that inadequate waste services lead to litter which does not only constitute visual pollution but also environmental degradation and health risks. Several projects are put in place to ensure the objectives of the programme are realised. These include the development of landfill sites, construction of waste transfer stations, construction of buy-back/recycling centres, construction material recovery facilities, composting facilities, street cleaning and beautification, domestic waste collection, greenest municipalities' competitions (GMC), and integrated waste management plans (IWMP). The following objectives are highlighted in the implementation of the programme:

- Promote environmental education and awareness to the communities especially as they are the main waste generators;

- Support the use of environmentally friendly waste disposal technology;
- Create sustainable livelihoods through recycling of waste (waste collection & minimisation); and
- Create and support mechanisms for the protection of environmental quality.

Eco-Furniture Programme

The idea of this programme is to capitalise on the untapped usefulness of invasive alien plant species (IAPS) in the manufacture of products in tandem with government needs, and in so doing, reducing the financial implication of clearing the invasive plants, maximising job creation and skills transfer opportunities. Since its inception, the programme has provided non-accredited and accredited training (registered under the Construction Education and Training Authority) to beneficiaries with a focus on life, business, and technical skills. The following objectives are highlighted in the implementation of the programme:

- Generate a value out of invasive alien plants to generate an additional income to clear more invasive plants;
- Utilise accessible and utilisable invasive alien plant biomass available in the areas to be harvested in the making of products;
- Create jobs in making value-added products relevant to the government's needs; and
- Produce high-quality products at costs lower than if the government buys them on the open market.

Biodiversity Economy Programme

South Africa, as the third most biologically diverse country in the world and with one of the largest natural capital assets, has identified biodiversity to be economically viable to economic well-being and a vehicle for the social upliftment of the citizens. By sustainably utilising South Africa's natural resources which are currently compromised, it is believed poverty can be reduced, leading to economic growth. The programme intends to focus on supply-side capacity, quality standards, certification, access to investment credit, benefit sharing and other support mechanisms for national and international bio-trade and green economy initiatives. To achieve the enormous benefits associated with this sector through the programme, the government aims to invest in biodiversity conservation and infrastructure development that enable the sustainable beneficiation and full participation of communities. With this, the vision is to have communities that support, uphold, and thrive from biodiversity conservation. The following objectives are highlighted in the implementation of the programme:

- Fair access and equitable sharing of benefits arising from bioprospecting involving indigenous biological resources promoted; and
- Sustainable utilisation and regulation of biological resources.

People and Parks Programme

The People and Parks Programme (P&PP) was borne out of the 2003 World Parks Congress held in Durban. This programme focuses on the people who are directly affected or threatened with removals from protected areas (PA) to make way for wildlife conservation. The programme identifies steps that address issues at the interface of communities and wildlife conservation while highlighting and implementing the rights of the communities affected by conservation processes. This has led to a P&PP with established community involvement, solid structures, frameworks, and policies to ensure the further success and sustainability of the programme. The goals of the programme are to ensure the promotion of biodiversity values in the proclaimed protected and surrounding areas, ensure that local communities are involved in the management of protected and surrounding areas, and invest in infrastructure development and biodiversity conservation for economic benefits. The following strategic objectives are highlighted in the implementation of the programme:

- Sustainable utilisation and regulation of biological resources;
- Promotion of fair access and equitable sharing of benefits from biological resources; and
- Conservation and protection of biodiversity and mitigation of threats.

Intergovernmental Platform on Biodiversity and Ecosystem Services

By committing to a strengthened science-policy interface for evidence-based decision-making, the Government of South Africa through the Department of Forestry, Fisheries, and the Environment established the Intergovernmental Science-Policy Platform on Biodiversity and Ecosystem Services (IPBES). The objective of the IPBES is to strengthen the science-policy interface for biodiversity and ecosystem services for the conservation and sustainable use of biodiversity, long-term human well-being, and SD. The establishment of IPBES is centred on four functions, namely knowledge generation (identify knowledge needs of policymakers and catalyse efforts to generate new knowledge), assessment (deliver global, regional and thematic assessments, and promote and catalyse support for sub-global assessment), policy support tools (identify policy-relevant tools/methodologies, facilitate their use, and promote and catalyse their further development), and capacity building (prioritise key capacity building needs and provide and call for financial and other support for priority needs). The following are the principles guiding the effectiveness of the programme:

- Contribution of indigenous and local knowledge;
- Scientific independence, credibility, relevance, and legitimacy;
- Policy-relevant but not policy-prescriptive;
- Full participation of developing countries;
- Collaboration – avoiding duplication;

- Gender equity;
- Inter- and multi-disciplinary approach; and
- Addressing terrestrial, marine, and inland water biodiversity and ecosystem services and their interactions.

Non-motorised Transport South Africa

Approximately 80% of South Africans depend on public means of transportation because the expansion of public transport and infrastructure for non-motorised transport (NMT) such as bicycle and pedestrian walkways has been neglected in cities. Inhabitants of poorly integrated townships spend a high percentage (30% or more) of their monthly expenses on transport costs; this becomes a major financial burden. The government through its departments, commits to promoting environmental, economic, and social development by implementing programmes such as the NMT.

The NMT plays a significant part in enhancing and complimenting existing public transport systems in the country as it provides secured access to public transport through cycling or walking (identified as healthy and cost-effective). The KfW Development Bank, on behalf of the German government, agreed to finance the construction of NMT in selected South African cities. This project allows citizens to access public facilities in a safe and eco-friendly manner. Apart from financing the project, the remaining duration of the project till the end of 2021 will focus on increasing the availability of bicycles and awareness creation.

Green Fund

The Green Fund was established by the Government of South Africa through the DEA to support the passage to a climate-resilient, resource-efficient, and low-carbon path which delivers high-impact social, environmental, and economic benefits. Kickstarting the programme, an initial allocation of R800 million was made available for disbursement by the Development Bank of South Africa which is the appointed implementing agent. This scheme is targeted solely at providing catalytic funds to facilitate investments in green initiatives that support job creation and poverty alleviation.

To address market weaknesses presently hampering the transition of South Africa to a green economy, the Green Fund aims to promote innovative and high-impact green programmes and projects; reinforce climate policy objectives through green interventions; build an evidence base for the expansion of the green economy; and attract additional resources to support South Africa's green economy development. Three main funding windows identified under the Green Fund programme to aid the transition to a green economy are the following:

- Green cities and towns (GCT) focusing on sustainable transport; sustainable waste management and recycling; renewable energy, including off-grid and mini-grid; sustainable water management; ecosystem services; energy efficiency and demand side management; sustainable human settlements; the built environment and GBs.

- Low carbon economy (LCE) focusing on energy efficiency; renewable energy; sustainable transport; biogas and biofuels; rural energy, including off-grid and mini-grid; and industrial cleaner production and consumption projects.
- Environmental and natural resource management (NRM) focusing on payment for ecosystem services (PES) projects; land use management and models; rural adaptation projects and plans; and biodiversity-benefiting businesses, including sustainable farming.

Greenest Municipality Competition

The Cleanest Town Competition (CTC) which was established in 2001, focuses on implementing the National Waste Management Strategy, which contains the 3Rs (reducing, recycling, and reusing waste materials) as key elements. Although the CTC programme was passably successful in achieving its objectives, incorporating the concept of SD has made it necessary to modify the programme, hence the new name, Greenest Municipality Competition (GMC). As an enhancement to the former CTC, the GMC is a linkage to other national and global initiatives such as greening the nation, reducing greenhouse gases (GHGs), and green goals amongst others. The GMC encompasses the following five major elements:

- Waste management;
- Energy efficiency and conservation;
- Water management;
- Landscaping, tree planting, and beautification;
- Public participation and community empowerment; and
- Leadership and institutional arrangements.

Greening and Open Space Management

Greening and Open Space Management are targeted at addressing poorly managed areas (such as eroded areas, areas overgrown with vegetation, illegal dump sites, and unmanaged open spaces) which not only create bad scenery but also cause health hazards, support criminal activities, and result in poor waste management. This programme proffers solutions to issues such as neglected open spaces in communities, illegal dumping sites, lack of social resources, lack of recreation for family and community gatherings, lack of empowerment for communities with historical indigenous information, and lack of social wellbeing that enhances healthy and strong communities. The following objectives are highlighted in the implementation of the programme:

- Encouraging the use of greener technologies to mitigate environmental degradation;
- Improving climate change adaptation through minimisation of biodiversity loss;
- Maximising measures towards pollution mitigation; and
- Restoring, enhancing, and rehabilitating open spaces.

Sustainability Assessment Tools for Buildings in South Africa

Compared to other countries on the continent of Africa, South Africa is at the fore-front of establishing, adopting, and implementing initiatives, policies, and frame-works towards the transition to a sustainable environment. Part of this conscious action is evidenced in the proliferation of GBs and SC practices in the country. There are single-criteria (such as Net Zero,) and multiple-criteria assessment tools (Green Star SA) presently utilised in South Africa. However, notable GBRTs or SATs used to evaluate the sustainability performance of buildings in South Africa include Green Star SA, Energy Water Performance (EWP) tool, Sustainable Build-ing Assessment Tool (SBAT) and Excellence in Design for Greater Efficiencies (EDGE). This section gives an overview of SBAT and Green Star SA. The former is borne out of research studies while the latter is an adaptation of existing tools by the Green Building Council South Africa for use in the country.

Sustainable Building Assessment Tool (SBAT)

According to Hill et al. (2002), the Sustainable Building Assessment Tool (SBAT) was developed by the Council for Scientific and Industrial Research (CSIR) as a response to the shortcomings of other GBRTs. The tool development spearheaded by Jeremy Gibberd draws on international best practices refined to the policies and local context of South Africa to provide a comprehensive methodology for as-sessing the sustainability performance of building projects (Sebake, 2008). SBAT comprises 15 objectives or evaluation categories that provide a simple and justifi-able effective assessment of the level of aiding the achievement of SD in building projects (Gibberd, 2003). It is, however, the only building assessment tool in South Africa and developing countries of Africa that attempts to incorporate the three dimensions of SD (environmental, economic, and social) in evaluating the sustain-ability performance of building projects.

SBAT can be employed to serve as a decision support tool for building pro-fessionals and stakeholders, as part of a brief to the building design team at the preconstruction stage, and as a medium of guaranteeing the integration and imple-mentation of sustainability policies into the CI. There are 15 categories within the SBAT with corresponding 73 criteria and about 74 indicators in total with which the tool evaluates the sustainability performance of new or existing buildings. The evaluation categories are Occupant Comfort; Inclusive Environments; Ac-cess to Facilities; Participation and Control; Education, Health, and Safety, Local Economy; Efficiency of Use; Adaptability and Flexibility; Ongoing Costs; Capital Costs; Water; Energy; Recycling and Reuse; Site; and Materials and Components. Table 8.2 presents the threshold categories and criteria for evaluation under the SBAT for building projects.

Green Star SA

In the past two decades, there have been efforts to develop voluntary standards to aid the adoption and implementation of GBs globally (Thatcher & Milner, 2012). To

Table 8.2 Sustainable Building Assessment Tool evaluation categories and criteria

Aspects	Categories	Criteria
Social Issues	Occupant Comfort	Lighting
		Ventilation
		Noise
		Views
		Access to green outside
	Inclusive Environments	Public transport
		Routes
		Changes in level
		Edges
		Toilets
	Access to Facilities	Childcare
		Banking
		Retail
		Communication
		Residential
	Participation and Control	Environmental control
		User adaptation
		Social spaces
		Amenity
		Community involvement
	Education, Health, and Safety	Education
		Security
		Health
		Smoking
		Safety
Economic Issues	Local Economy	Local contractors
		Local building material supply
		Local component manufacturer (furniture?)
		Outsource opportunities
		Repairs and maintenance
	Efficiency of Use	Useable space
		Occupancy
		Space use
		Use of technology
		Space management
	Adaptability and Flexibility	Vertical dimension
		Internal partitions
		Services
	Ongoing Costs	Maintenance
		Cleaning
		Security/care taking
		Insurance/water/energy/sewerage
		Disruption and 'downtime'
	Capital Costs	Consultant fees
		Buildability
		Construction
		Shared costs
		Sharing arrangements

(*Continued*)

Table 8.2 (Continued)

Aspects	Categories	Criteria
Environmental Issues	Water	Rainwater
		Water use
		Grey water
		Runoff
		Planting
	Energy	Location
		Ventilation system
		Heating and cooling system
		Appliances and fittings
		Renewable energy
	Recycling and Reuse	Toxic waste
		Inorganic waste
		Organic waste
		Sewerage
		Construction waste
	Site	Brownfield site
		Neighbouring buildings
		Vegetation
		Habitat
		Landscape inputs
	Materials and Components	Embodied energy
		Material/component sources
		Manufacturing processes
		Recycled/reused materials and components
		Construction processes

Source: Author's compilation (2019).

align with the international GB movement, the Green Star SA rating tool was developed by the Green Building Council South Africa (GBCSA) and duly recognised by the South African government. Adapted from the Green Star system by the Green Building Council of Australia (GBCA), the developed Green Star SA certification identifies and considers the social, economic, and climatic context of South Africa (Mavhungu, 2019). The development is aimed at providing an objective GB evaluation in South Africa and Africa while recognising and rewarding environmental leadership in the CI (GBCSA, 2017a). The Green Star SA rating tool was developed to establish a common language and standard of measurement of GBs; promote integrated, whole-building design; identify building lifecycle impacts; raise awareness of green building benefits; recognise environmental leadership; and transform the built environment to reduce the environmental impact (GBCSA, 2014).

There are tools within the Green Star SA for accessing and certifying various building types and interiors (Van Reenen, 2014). Tools for Existing Building Performance (EBP) focuses on the operational phase of a building's lifecycle, Interiors evaluates the environmental characteristics of any buildings' interior fit-outs, New Buildings and Major Refurbishments focuses on design features at the tender or construction stage of building projects, Sustainable Urban Precincts assesses the environmental performance of the different development phases of precincts and neighbourhoods, Green Star Custom addresses building types that do not fit into the existing tools, and Green Star Africa which serves as a natural touch point for GB movements and councils in other parts of Africa (GBCSA, 2017a).

The Green Star SA rating tool identifies and rewards initiatives that reduce the environmental footprint of building projects (Goosen, 2009) under ten categories of Management, Indoor Environment Quality, Energy, Transport, Water, Materials, Land Use and Ecology, Emissions, Innovation, and Socio-Economic. The categories, indicators and corresponding credits for the Green Star SA Existing Building Performance are presented in Table 8.3. Based on the total credits earned, the following

Table 8.3 Green Star SA evaluation categories, criteria, and rating benchmarks

Green Star SA evaluation categories and criteria

Categories	Criteria
Management	Certified buildings
	Accredited professional
	Building management
	Green cleaning performance
	Green leasing
	Ongoing monitoring and metering
Indoor Environmental Quality	Indoor air quality
	Lighting comfort
	Thermal comfort
	Occupant comfort survey
	Acoustic quality
	Daylight and views
	Indoor pollutant management
Energy	Energy consumption (greenhouse gas emissions)
	Peak electricity demand
Transport	Alternative transportation
Water	Portable water
Materials	Procurement and purchasing
	Solid waste management
Land Use and Ecology	Ecological and site management
	Grounds-keeping practices
Emissions	Refrigerants
	Legionella
	Storm water
Innovation	Innovative strategies and technologies
	Exceeding green star SA benchmarks
	Environmental initiatives

Green Star SA rating benchmarks

Rating	Overall Score	Remarks
One Star	10–19	Committed to Performance
Two Star	20–29	Committed to Performance
Three Star	30–44	Committed to Performance
Four Star	45–59	Best Practice
Five Star	60–74	South Africa Excellence
Six Star	75+	World Leadership

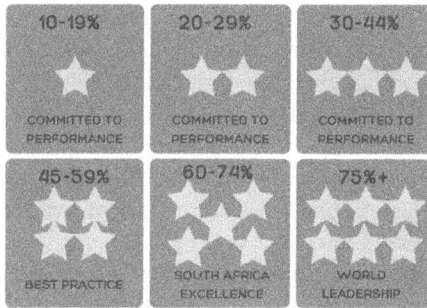

Figure 8.2 Green Star SA recognition and certification levels

levels of certification and recognition are awarded to the building project under assessment: Acknowledgement, Best Practice, South Africa Excellence, and World Leadership as shown in Figure 8.2. However, the score for each evaluation category is determined based on the percentage of the achieved credit as shown below:

$$\text{Category Score} = \frac{\text{Number of points achieved}}{\text{Number of points available}}$$

Green Building Collaborative Networks in South Africa

There are numerous non-governmental organisations in South Africa at the forefront, clamouring for the holistic adoption and implementation of SD in the country. These organisations are also complemented by various professional bodies and associations in the country with specific arms and committees dedicated to incorporating elements of sustainability in their processes, activities, and operations. This section discusses the Sustainability Institute (SI), a training, research and consulting entity, and the Green Building Council South Africa (GBCSA) which administers the Green Star SA rating tool, offers training for assessors and is at the frontline of creating awareness of the benefits of GBs.

Sustainability Institute South Africa

Established in 1991, the Sustainability Institute (SI) in South Africa provides a space for people to explore approaches to creating a more sustainable society (SI, 2020). The Institute offer master's and PhD programmes in SD (in partnership with the University of Stellenbosch School of Public Leadership) as well as research and consulting services. Since 2011, the SI has been involved in research contract work, offering its clients access to sustainability expertise in areas such as Alternative Economics, Social Entrepreneurship, Sustainable Food Systems, Corporate Sustainability and Governance, Renewable Energy, and Cities and Infrastructure. The core areas of services provided by the institute include Sustainability Strategy, Capacity Building, Research and Authorship.

Located in Lynedoch Ecovillage, the SI also focuses on impacting the host community which led to the establishment of the Lynedoch Children House, and aftercare programmes with an extended partnership with SPARK Schools, also in Lynedoch. From donations and gifts received, the Institute is also able to ensure indigenous children and young adults can attend the Lynedoch Children's House, benefit from the Lynedoch Youth Hub that provides strong academic support, participate in swimming lessons under the swimming project of the Institute, join the Lynedoch United Football Club which helps grow young talents in the community, participate in the Karate Programme for learners of SPARK Lynedoch, and join the Knitting Club which provides essential support and community to the elderly.

Green Building Council South Africa

The Green Building Council South Africa (GBCSA) is the most active national arm on the continent of Africa and one of the 75 members of the World Green Building Council (WGBC). The Council is made up of members who are passionate and collaborative individuals across the public and private sectors of the built environment with the intent to have buildings designed, built, and operated in an eco-friendly manner (GBCSA, 2017b). According to the WGBC (2020), the GBCSA was established in 2007 to lead the sustainability agenda within the built environment of South Africa while providing networks, knowledge, training, and tools to promote GB practices across the country.

The modus operandi of the GBCSA revolves around four major areas of advocacy, education, certification, and membership. The advocacy part entails the various mediums through which the Council create awareness of the processes, strategies, and benefits of adopting and implementing GBs in South Africa. These include Planet Shapers, +Impact magazine, events, an annual convention, students' annual Greenovate competition, campaigns, and initiatives. The education part is the educative platform to provide professional development for professionals in the built environment focusing on GB, Green Star, EDGE, Net Zero and EWP rating tools. This platform performs its objectives through accredited professional programmes, workshops, online courses, and online exams. The certification part which is administered by the Council involves building evaluation using the existing rating tools (Net Zero, Green Star, EDGE, and EWP) to provide an objective and credible GB assessment in South Africa and Africa. The membership category involves the offering of individuals, organisations and companies the opportunity to join the GB movement while enjoying a series of benefits (discounts, knowledge, high-level networking, and increased profile) and at the same time playing their part in moving South Africa to a sustainable state.

Summary

This chapter provided an overview of sustainability policies, drives, initiatives, and frameworks in South Africa. The notable green building collaborative networks in South Africa are also presented in this chapter. An analysis of the existing green building rating tools and sustainability assessment tools (SATs) used in South Africa is also presented in this section while providing an understanding of the various evaluation categories and criteria that constitute these tools. The next chapter reviews the various gaps in SAT research to achieve the objective of this book. The root of these gaps in nature is also presented and discussed thereby providing the biomimicry thinking lens underpinning the conceptualised tool.

References

Department of Environmental Affairs and Tourism (2008). *A national framework for sustainable development in South Africa.* South Africa: Department of Environment, Forestry and Fisheries.

Department of Forestry, Fisheries and the Environment. (2019). *Sustainable development.* Retrieved 24 March 2020 from https://www.dffe.gov.za/projectsprogrammes

Gibberd, J. (2003). *Building systems to support sustainable development in developing countries.* Pretoria, South Africa: CSIR Building and Construction Technology.

Goosen, H. J. H. (2009). *Green star rating, is it pain or glory?* [BSc Hons Mini-dissertation, University of Pretoria, South Africa].

Government of South Africa. (2020). *National Development Plan 2030.* Retrieved 24 March 2020 from https://www.gov.za/issues/national-development-plan-2030

Green Building Council South Africa (2014). *Technical manual: Green Star SA existing building performance pilot.* South Africa: Green Building Council South Africa.

Green Building Council South Africa. (2017a). *Green Star certification.* Retrieved 22 November 2017 from https://gbcsa.org.za/certify/green-star-sa/

Green Building Council South Africa. (2017b). *About GBCSA.* Retrieved 24 March 2020 from https://gbcsa.org.za/about-gbcsa/

Green Growth Knowledge Platform. (2020). *South Africa National Development Plan 2030: Our future – make it work.* Retrieved 24 March 2020 from https://www.greengrowthknowledge.org/national-documents/south-africa-national-development-plan-2030-our-future-make-it-work

Gunnell, K. (2009). *Green building in South Africa: Emerging trends.* South Africa: Directorate: Information Management, Department of Environmental Affairs and Tourism (DEAT).

Gupta, K., & Laubscher, R. F. (2018). South African Government initiatives toward a transition to green economy. In M. Yülek (Ed.), *Industrial policy and sustainable growth. Sustainable development.* Singapore: Springer.

Hill, R., Bowen, P., & Opperman, L. (2002). Sustainable building assessment methods in South Africa: An agenda for research. *Sustainable building* (pp. 1–6). Rotterdam (Netherlands): In-house publishing.

James, P. M., Rust, A. B., & Kingma, L. (2012). The well-being of workers in the South African construction industry: A model for employment assistance. *African Journal of Business Management, 6*(4), 1553–1558.

Mavhungu, Z. S. (2019). *Determination of the effect of Green Star certification office rental in Sandton, South Africa.* Johannesburg, South Africa: University of the Witwatersrand.

National Planning Commission. (2020). *The National Development Plan.* Retrieved 24 March 2020 from https://nationalplanningcommission.wordpress.com/the-national-development-plan/

Ross, N., Bowen, P. A., & Lincoln, D. (2010). Sustainable housing for low-income communities: Lessons for South Africa in local and other developing world cases. *Construction Management and Economics, 28*(5), 433–449.

Sebake, T. N. (2008). Review of appropriateness of international environmental assessment tools for a developing country. *World Sustainable Building Conference*, 1–7.

Sustainability Institute. (2020). *Who we are*. Retrieved 24 March 2020 from https://www.sustainabilityinstitute.net/about/who-we-are

Thatcher, A., & Milner, K. (2012). The impact of a 'green' building on employees' physical and psychological wellbeing. *Work (Reading, Mass.)*, *41*(*Suppl. 1*), 3816–3823.

United Nations. (2004). *Johannesburg plan of implementation*. Retrieved 24 March 2020 from https://www.un.org/esa/sustdev/documents/WSSD_POI_PD/English/POIChapter8.htm

United Nations. (2020). *The National Framework for Sustainable Development*. Retrieved 24 March 2020 from https://sustainabledevelopment.un.org/index.php?page=view&type=99&nr=172&menu=1449

Van Reenen, C. A. (2014). *Principles of material choice with reference to the Green Star SA rating system*. SA: Alive2Green.

Windapo, A. O. (2014). Examination of green building drivers in the South African construction industry: Economics versus ecology. *Sustainability*, *6*(9), 6088–6106.

Windapo, A. O., & Cattell, K. (2013). The South African construction industry: Perceptions of key challenges facing its performance, development, and growth. *Journal of Construction in Developing Countries*, *18*(2), 65–79.

World Green Building Council. (2020). *Green Building Council South Africa*. Retrieved 24 March 2020 from https://www.worldgbc.org/member-directory/green-building-council-south-africa

Zizzamia, R. (2020). Is employment a panacea for poverty? A mixed-methods investigation of employment decisions in South Africa. *World Development*, *130*, 1–15.

Part IV

Biomimicry Sustainability Assessment Tool (BioSAT)

9 Gaps in Sustainability Assessment Tools Research

Introduction

Despite the global clamour for the adoption and implementation of sustainability, Africa as a continent is yet to consciously embrace this new paradigm. To address the global sustainability issues, the implementation and enforcement of codes, legislations, and tools are identified as a working approach by authorities to ensure a healthy and safe built environment (Van der Heijden, 2017). An examination of the existing green building rating tools (GBRTs) or sustainability assessment tools (SATs) mostly in use globally with none tailored for Africa is a pointer to the fact that the continent is lagging in making significant efforts geared towards sustainable development (SD) drives. Several sources of literature have confirmed that most African countries do not have any policy, framework, or legislation aligned to SD while their construction professionals are likewise unaware of sustainable construction principles (SCPs). According to the World Green Building Council (WGBC) which is one of the organisations advocating for the adoption of green buildings (GBs) globally, South Africa is the only country with an established national GB council. Apart from South Africa, Cameroon, Egypt, Ghana, Kenya, Mauritius, Morocco, Namibia, Rwanda, and Tanzania are the only countries in Africa with a prospective membership status of the WGBC (WGBC, 2020). The critical reaction mechanism for countries to meet their global commitments of addressing carbon emissions and other environmental challenges (Hydes et al., 2018) becomes an uphill task when such a country lacks the presence of a body like the GB Council, a phenomenon common in Africa.

Another major observation from most research works on GBRTs, and SATs revealed the paucity of rating tools tailored to the sustainability needs of a region or geographic location. Until recently when researchers, academic institutions, professional bodies, organisations, and governmental agencies nationally embarked on drives, collaborations, and efforts at either adapting notable assessment tools such as BREEAM, LEED, Green Globes, Green Star, and CASBEE or developing one that addresses their needs, the proliferation of GBs remain stagnated on the continent of Africa. While most of the current rating tools are made up of different categories, an important and underlying factor such as a comprehensive approach to sustainability (Say & Wood, 2008) is lacking in the majority of the GBRTs

DOI: 10.1201/9781003415961-13

utilised and adopted globally. If a tool addresses only the environmental aspect of sustainability or SD, how does it help the drive towards achieving holistic sustainability when equally significant themes such as economic and social are left out?

Hence, the GBRTs and SATs used in developed and other advanced countries cannot be wholly adopted and integrated for use in Africa owing to certain situations and peculiarities. These peculiarities are factors around cultural, tribal, economic, social, political, and climatic conditions in Africa, hence the need for the development of an SAT that addresses and considers these issues. To address the gaps, biomimicry, which is the study and emulation of outstanding phenomena in nature to provide sustainable solutions to human challenges, is adopted. Questing for holistically sustainable methodologies, scientists, researchers, and professionals now consult and learn from nature's experience from 3.8 billion years of evolution during which it has mastered what works and what does not (Benyus, 2011). According to Hargroves and Smith (2006), this novel field has provided timely and significant innovative solutions in areas such as water reuse and energy engineering where improvements in multiple-scale efficiency are required. As further affirmed by Folke et al. (2011), a deliberate shift from nature and people as disjointed parts to interdependent and connected systems (reconnecting to nature) will help achieve the global sustainability objectives owing to the opportunities it provides for social and economic development.

Gaps in Sustainability Assessment Tools Conceptual Framework

This section of the study identifies the gaps in the sustainable assessment tool (SAT) conceptual framework. According to Adom et al. (2018), a conceptual framework explains the researcher's description of the relationship between the key ideas of a study and how the research problem of the study will be examined. Since a conceptual framework can also be the adaptation of variables from existing theories to suit a researcher's study purpose, it is imperative to identify the gaps that existing GBRTs and SATs have failed to address in the assessment of buildings in Africa. This section of the study attempts to address the three gaps that have been identified, namely resilience, inclusivity and equality, and collaboration. The consideration of these gaps is premised on the notion that a truly sustainable building cannot be achieved by its compliance with only one part of the three sustainability dimensions. Based on the SD concepts of the triple bottom line (economy, social, and environment) which describes their interdependencies and interrelationships, it is assumed that each dimension is equally important and should not be ignored (Romero-Lankao et al., 2016). Despite being multiple-attribute rating tools, a simple perusal of most of the existing and notable GBRTs and SATs used and adopted globally revealed they are only evaluating a part or portion of the three dimensions of sustainability. While some of these assessment tools are constantly updating and upgrading, those which tend to address sustainability issues holistically failed to consider country-specific factors, especially as they relate to the African continent. By comprehensively observing the 17 Sustainable Development Goals (SDGs), the gaps in the existing SATs are laid bare. Hence, the clamour by researchers and sustainability proponents that the development of GBRTs and SATs must take into cognisance, amongst many others, country-specific factors such as the culture of the people while others agree that these assessment tools should be focused on people. This is because each country experiences different and specific challenges in its quest to achieve SD while also admitting the fact that African countries face unique problems compared to other continents of the world (UN, 2015).

Gap One: Resilience

As established in the literature, the successful implementation of SD is dependent on how it is adapted and tailored to the needs and peculiarities of each country. With Africa always lagging in embracing and adopting global technological trends and standards, it is imperative to ensure that any development and policy takes into consideration factors that are country-specific. As supported by Markelj et al. (2014), the development, adaptation, and use of SATs must consider local rules, legislations, languages, and systems of measurement, coupled with climatic, cultural, geographical, political, and economic situations in the country or region. According to Scott (2017), the process of globalisation, climate change, and natural disasters are some of the factors that pose grave threats and increasing disturbances to the built environment unless elements of resilience are incorporated into buildings. A skeletal perusal of the existing GBRTs and SATs as discussed in previous chapters of this book revealed the absence of resilience among the categories, criteria, and sub-criteria. Hence, the need to evaluate resilience in buildings is one of the key parameters to ascertain their sustainability performance.

There have been several discussions around the concept and research models of resilience theory. As a subject that transcends disciplines and sectors, resilience theory constitutes different and sometimes confusing theoretical and conceptual underpinnings. However, the intent when dissected properly tends towards the same description as inferred in the diverse definitions of resilience. According to the study of Ledesma (2014) on leadership, thriving, recovery, and survival are the three concepts around the theory of resilience while also describing the state in which a person (system) may be during or after a disturbance. The focus of resilience theory is on positive individual, social, and contextual variables that disrupt or affect developmental routes (Zimmerman, 2013). It is, therefore, crucial to note that unless resilience theories are tested, the concept of resilience itself is devoid of genuine content or utility (Shean, 2015).

Resilience is a debatable concept when tracking its origination, meaning, and application as there are genealogical offerings drawn from the international relations, psychology, political science, disaster management, engineering, social science, and ecology fields (Rogers, 2015). The subject of resilience has become a definite policy intention for equitable and SD (Bousquet et al., 2016). Resilience first appeared in literature regarding material science in the 1800s, gained popularity in political policy and discourse and only appeared in ecological literature when Holling applied it in the 1970s (Van Wyk, 2015). The recent global experience of and actions on climate change and natural disasters have therefore made the subject of resilience an imperative one. Given the lack of an agreed definition among scholars and practitioners since the subject of resilience is used in different sectors and fields, the clarification of the definition used in the study is important (Thiede, 2016). Arguably, the concept of resilience is more established in ecology while resilience thinking proffers a notable epistemic challenge to the existing ecological knowledge base (Evans & Reid, 2013; Walsh-Dilley & Wolford, 2015).

As described by Holling (1973), resilience is the 'measure of the persistence of systems and of their ability to absorb change and disturbance and still maintain the same relationships between populations or state variables'. In the book titled 'Design for Environment: A Guide to Sustainable Product Development', resilience is described as the ability of an enterprise to grow, adapt, and survive in tempestuous situations (Fiksel, 2009). In the study of Rademaker et al. (2018) on urban resilience, it is defined as the potentiality of a city in the event of a destructive (climate-related) situation to retain during, recuperate from, and attune to variations in the system. Within the fields of engineering, physics, and mathematics, Romero-Lankao et al. (2016) define resilience as the adaptive ability of an ecosystem to recover and return to a stable state after any disturbance, such as the restoration and recovery of a template forest after floods. Based on the numerous ecological perspectives of resilience, Rademaker et al. (2018) opined that in the case of cities that do not only change when certain thresholds are reached, resilience is better applicable to non-ecological subjects.

Resilience is identified as a vital feature in a difficult, connected, and unpredictable world like ours, as it allows the duo of human and biological systems to successfully survive a persistent torrent of change (Fiksel, 2009). However, based on the numerous definitions put forward by researchers and authors, Carlson et al. (2012) defined resilience as the capacity of a system (region, community, organisation, asset) to anticipate, resist, absorb, respond to, adapt to, and recover from a disaster or disturbance. This definition is further described in Figure 9.1 as adapted from the study of Carlson et al. (2012) showing components of resilience and the timing of an adverse event in a system. Resilience is resilient as a concept and is

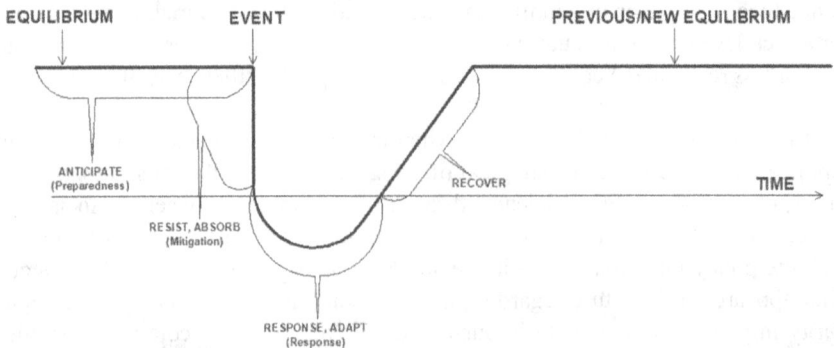

Figure 9.1 Components of resilience and timing of a disaster in a system

immensely expandable and adaptable symbolising ingenuity, endurance, collective intelligence, activation, adaptability, expansion, democratic mutualism, and preparedness (Vrasti & Michelsen, 2017). Adaptability and complexity are also identified by Desjardins et al. (2015) when determining what makes a system resilient. These characteristics are what enable a resilient system to achieve its goal of responding to evolving situations while ensuring the system and its essential functions experience minimal losses (Redman, 2014).

To achieve or build resilience, there are approaches identified as effective and efficient. As identified and explained by the study of Langeland et al. (2016), impact mitigation, real-time adaptation or response, and recovery efforts are advanced approaches to building resilience. Impact mitigation describes the ability (absorbing damages and minimising lost capacity) of a system to maintain its operational capability in the event of a disaster. Real-time adaptation or response describes the ability of a system to remain operational (by relying on easily replaceable sub-capabilities and flexibility) in the event of a disaster by adjusting or adapting its operations. Lastly, recovery efforts describe the activities (operating through minor adaptations and ensuring ample and rapid recovery) that enable a system to regain mission-vital capabilities after a disaster or impact. The study of Rademaker et al. (2018) on urban resilience in a similar form proffers maintaining, recovering, and learning as the three approaches and dimensions of resilience. The maintaining dimension of resilience entails the capacity of a system to survive the impact of a climate event or disturbance, withstand material degeneration, and sustain the operation of the urban domain. The recovering aspect describes the rate at which a system can return to its normal operation after experiencing the impact of a climate event or disturbance. The learning dimension describes the ability to interpret lessons from previous occurrences into substantial and workable policies and act accordingly upon the policies in a contextual manner. The study of Folke et al. (2010) indicated that resilience thinking in a system embodied three aspects, namely transformability, adaptability, and persistence. Transformability entails the capacity of a system to establish new stable domains for development and cross thresholds into new development trajectories and new stable landscapes. Adaptability describes the capacity of a system to adjust and adapt the system's responses to changing internal and external drivers and processes while allowing for growth within the current stable domain along the current path. Lastly, persistence describes the capacity of a system to change and remain (continuously change and adapt yet remain) within a stable state in critical thresholds. However, it is imperative to link resilience with sustainability to enable the plan and design considerations for resilience to achieve the desired sustainable systems (McPhearson, 2014).

From the widely accepted Brundtland's definition, SD refers to developments that address present needs without disabling future generations' ability to meet their needs. Premised on the need to achieve maximum good for human society and the environment, there have been several attempts by researchers around the world to integrate resilience with the theoretical approaches and concepts of sustainability (Redman, 2014). According to van Wyk (2015), a precondition for sustainability is the adaptive capability of resilient systems (including socioeconomic and biological entities) which may lead to new equilibria. As affirmed by Stevens and Schieb (2007), reliable and resilient infrastructure is one of the necessary traits for a sustainable society. It is therefore important to integrate the concepts of sustainability and resilience in a complementary manner. The study on coupled human-environment systems (CHES) by Turner (2010) indicated that sustainability science helps in uncovering the attributes within the human and environmental systems that make them more resilient (less vulnerable) to perturbations, stressors, disturbances, disasters, and the like.

In the process of identifying the ways by which the concepts of resilience and sustainability are integrated for implementation, three frameworks are inferred. According to the study of Marchese et al. (2018), the first framework describes resilience as a component of sustainability; the second framework describes sustainability as a component of resilience while the third framework describes resilience and sustainability as different conceptual objectives. The first framework describes resilience as a vital part of the sustainability concept when achieving sustainability is the main goal of a system. This framework maintains that a system is more sustainable by increasing its resilience; however, the system does not necessarily become more resilient by increasing its sustainability. The second framework describes sustainability as a vital factor to consider when achieving resilience is the main objective of a system. This framework maintains that a system is more resilient by increasing the sustainability of the system; however, the system does not necessarily become more sustainable by increasing its resilience. The third framework describes resilience and sustainability as concepts (with different objectives) that can compete with or complement each other. This framework maintains that sustainability does not contribute to resilience, nor does resilience contribute to sustainability, but that both concepts are affected individually or collectively by policies and projects. Hence, the study of Achour et al. (2015) posited that the combination of resilience with the three pillars of SD will result in infrastructures that are socially, economically, and environmentally efficient as well as resilient enough to withstand disasters or disruptions. As adapted from the study of Achour et al. (2015), the integration of resilience with sustainability (environment, social, and economic) dimensions is shown in Figure 9.2.

Biomimicry Thinking: Resilience Strategies in Nature

There has been an increased collaborative effort among biomimics, nature conservationists, and ecologists to mitigate, reverse, or slow the pressure exerted on the earth's life support systems due to the negative impact of unsustainable human activities (Power & Chapin, 2010). It is to this end that professionals, designers, scientists, and innovators around the world are now consulting and tapping from the vast deposit of ideas and information in the natural world. According to Habib and Nagata (2019), organisms in biological systems possess and exhibit outstanding attributes (systems, algorithms, processes, and structures) that enable them to perform (defence, mobility, and functionality) in a sustainable manner. For example, a platypus can spot prey via an integrated sensitivity to touch, smell,

Figure 9.2 Integration of resilience with sustainability dimensions

and electroreception; a worm relies on its sensitive skin, chemical receptors, and magnetoreception to evade predators, discover food, and navigate direction (Liu et al., 2019). The functional data drive in biological fields has therefore, supported biomimicry as one of the triple Bs (Big data, Biotechnology, and Biomimicry) with the potential to enhance resilience and sustainability (Allam, 2020).

The study of Walker et al. (1999) indicated and emphasised the imperativeness of functional diversity in natural organisms as key to ensuring their resilience and that of their ecosystem under changing environmental conditions or disturbances. Juxtaposing this inference within the built environment is believed to have the potential to ensure the holistic sustainability of buildings. As suggested by the study of Anderies et al. (2013) on resilience, sustainability, and robustness, firms (systems) should extensively engage in strategies that optimise the efficiency of risk management and natural resources. According to Galaz (2005), an underlying challenge facing humanity today is the management (learning and adaptation) of systems in environmental changes and disturbances (such as the unanticipated rise in the level of toxic pollutants and sudden flooding). Comprehending and optimising the properties of a system makes such a system more resilient to disturbances, disasters, and unidentified risks bearing in mind that resilience improves risk management practices (Böggemann & Both, 2014). Focusing on nature has therefore been identified as one of the most efficient avenues to address human issues. Perusing the fauna, flora, biotas, biomes, and the natural world (ecosystem) has proven to reveal outstanding innovative solutions to human challenges and methodologies for improving the efficiency and sustainability of what has been created. To study, emulate and infer resilience as exhibited by natural organisms, it is important to focus on those strategies, processes, forms, and attributes that have aided their sustenance through the over 3.8 billion years of nature's evolution. As informed by Olsson et al. (2014), resilience focuses primarily on complex socio-ecological systems where the emphasis is on human society and nature connectedness. Hence, the imperativeness of understanding the benefits and contributions of nature (eco-system services) to the human environment.

Ecosystem services (ES) originate from comprehensive natural cycles which are powered by solar energy, constituting the nature mechanism (Daily, 2003). They are perceived as the direct and indirect benefits obtained from or provided by nature. As described by Fisher et al. (2009), ES result from involute intercon-nection among organisms and their environment, complex utilisation paradigms, and diverse beneficiary perceptions. Despite the several definitions found in the literature, one of the Millennium Ecosystem Assessments remains widely refer-enced and acceptable. According to the study of Saito and Ryu (2020), ES is simply defined by the Millennium Ecosystem Assessment (2005) as benefits derived from nature (ecosystems) which include provisioning services (such as water and food); regulating services (such as control of diseases, drought, and floods); supporting services (such as nutrient cycling and soil formation); and cultural services (such as spiritual, recreational, and other nonmaterial benefits).

This definition is also corroborated by the study of Chan et al. (2006) which states that the essence of ES is to provide society with benefits from nature (eco-system). However, on the other hand, it identified seven benefits, namely water provision, forage production, biodiversity, flood control, crop pollination, outdoor recreation, and carbon storage. The study of Elmqvist et al. (2015) on urban ES maintains that ES is beneficial (monetary and nonmonetary) to human well-being and society at large, contributes to the maintenance of biodiversity, and aids the development of more resilient urban areas (systems). The study of Elmqvist et al. (2015) further identified six benefits provided by ES (green and blue spaces) to be the following: carbon storage and sequestration; pollution and air quality regula-tion; energy savings and temperature regulation; stormwater reduction; positive health effects; and recreation and other amenity services. These benefits and many more provided by ES are because of the numerous outstanding attributes of organ-isms in nature which are all optimised interdependently at the ecosystem level (holistic level of biomimicry), while in turn aiding their resiliency.

According to the study of Kunz et al. (2011) on bats, organisms in nature sig-nificantly contribute to the health and efficient functioning of ecosystems while supporting cogent regulatory functions such as erosion control, water filtration, nu-trient cycling, and climate regulation, amongst others. The cooperative functionali-ties in natural organisms exhibited via their shapes, processes, and operations have been identified as factors responsible for their longevity, efficiency, and resilience in the event of disturbances and disasters (whether natural or man-made). With the human history of nature exploitation and disruption, biomimicry, and sustainabil-ity proponents believe that the natural ecosystem will offer strategies and insight into sustainable solutions to the global challenges facing humanity focussed on and visualised as a source of inspiration. Regarding resilience, a keen observa-tion of activities and functions in the natural ecosystem has the potential to reveal countless lessons and insights for achieving resilience in buildings. Table 9.1 pre-sents a few of the resilience lessons, insights, and inspirations provided by nature as documented by AskNature which is the database for nature-inspired strategies,

Table 9.1 Ecosystem strategies for resilience

S/N	Function	Nature inspiration	Strategy
1	Manage compression, manage shear, manage tension	Scots pine, bones	Trunks and branches of trees withstand stresses through load-adaptive growth
2	Distribute liquids, optimise shape/materials, optimises water and nutrient flow	Lemon leaves, leaves	Vein systems in leaves allow for optimal flow and resilience to damage due to a dense network of nested, interconnected loops
3	Optimise shape/materials, prevent deformation	Biological tissues created out of numerous cells joined together	Biological tissues are created out of many cells joined together in a specific sequence to produce a high-performance tissue. When one cell is damaged, it can be easily replaced without function loss or disruption. Example is pneumocell architecture (low cost, lightweight, flexible, non-emitting and resilient inflatable structures)
4	Regulate water storage, maintain biodiversity, manage disturbance in a community, cycle nutrients, control erosion and sediment, detoxification/purification of air, water and waste, generate soil/renew fertility, and biological control of populations, pests, and diseases	Structures and functions in natural ecosystem	Mimicking natural ecosystems, a strategy that takes advantage of benefits found in natural systems such as self-regulation, biodiversity, food production, nutrient cycling, carbon sequestration, accumulation of ecological capital, and resilience to most perturbations. Example is permaculture, or perennial grain cropping
5	Cycle nutrients, control erosion and sediment, generate soil/renew fertility	Tallgrass prairie	Prairie ecosystems demonstrates sustainability as they maintain soil and water quality, and nutrient cycling due to their diversity and perennialism
6	Manage disturbance in a community, protect animals	Ecosystems	Ecosystems survive biotic and abiotic disturbances by having multiple species that respond in different ways
7	Maintain biodiversity, manage disturbance in a community	Prairie grassland	Diverse plant species in prairie grasslands support a long-term, stable ecosystem because they exhibit complementary functionality
8	Respond to signals, navigate through liquid, coordinate by self-organisation	Mosquitofish	Individual fish avoid contact as they move synchronously in shoal by reacting dynamically to the nearest neighbour fish

(Continued)

Table 9.1 (Continued)

S/N	Function	Nature inspiration	Strategy
9	Distribute energy, manage compression, manage impact, protect from animals, prevent fracture/rupture	Scaly-foot snail	The shell of the golden-scale snail protects from attack with a specialised tri-layered composition
10	Prevent fracture/rupture	Buriti palm	A spongy, lightweight, biodegradable foam produced by buriti palms remains elastic, rather than brittle at very low temperatures
11	Attach permanently, manages stress	Human-shoulder joint	Rotator cuff in humans manages stress (force) via an extra pliable region
12	Manage compression, manage impact, prevent fracture/rupture	Macadamia	The outer shells of the nuts of macadamia trees are extremely resistant to cracking due to their fullerene-like surface structure
13	Modify size, shape, mass, and volume	European hornbeam	Leaves of the hornbeam tree have a corrugated fold pattern that allows for a balance between rigidity and flexibility
14	Prevent buckling	Waterlilies	The ribbed underside of the Amazon water lily provides structural support to keep the leaf afloat and sustain small loads
15	Manage compression	Watercress	Tetra decahedral-shaped cells allow for maximal packing space while also resisting compression
16	Manage compression, manage impact, manage shear, manage tension, physically assemble structure	Plants	Cellulose fibres in plant stems increase toughness by winding around tubes at an angle.
17	Protect from excess liquids, attach temporarily, move in/on solids, protect from dirt/solids	Tokay gecko	Feet of the tokay gecko dynamically self-clean by flinging contaminants off toes
18	Differentiate signal from noise	Elephant fishes, South American knifefishes	Receptor organs in the skin of weakly electric fishes filter out background noise and enable communication using cells as capacitors
19	Sense shape and pattern in a living system, navigate through liquid, capture, absorb, or filter organisms, navigate through liquid, sense electricity/magnetisms from the environment	Elephantnose fish	The elongated chin on the elephantnose fish helps it hunt in murky water by sensing electric signals

(*Continued*)

Table 9.1 (Continued)

S/N	Function	Nature inspiration	Strategy
20	Protect from microbes, regulate reproduction or growth	North African blue tit	Blue tit females protect their chicks from pathogenic bacteria by selectively placing fragments of certain aromatic plants in their nests
21	Protect from fungi, protect from microbes, protect from excess liquids, protect from dirt/solids	Springtails, Springtail	Skin of springtails repels liquids with undercut structures that stabilise air bubbles
22	Protect from temperature, capture, absorb, or filter energy	Polar bear	Guards hairs on the polar bear prevent heat loss by absorbing heat in the form of infrared radiation
23	Distribute gases, maintain homeostasis, modify pressure, protect from temperature, optimise shape/materials	Leafcutter ant	Nests of leaf-cutting ants self-ventilate due to two different types of turrets that take advantage of wind
24	Transform mechanical energy, disperse seeds, distribute solids, move in/through gases	Japanese maple	The leading edge of the tornado-like spinning maple seed provides a constant lift force, allowing seeds to disperse over a greater area
25	Optimise shape/materials, store gases, expel gases, distribute gases	Birds	The respiratory system of birds facilitates efficient exchange of carbon dioxide and oxygen via continuous unidirectional airflow and air sacs
26	Manage compression, manage shear, manage tension, prevent buckling, prevent fracture/rupture	Venus flower basket	Each hierarchical level of organisation of silica skeleton in the Venus' flower basket sea sponge is tough and stable (yet flexible), helping to manage forces
27	Distribute liquids, protect from excess liquids	Poplar spiral gall aphid	Powdery wax secreted by aphids makes sticky honeydew waste manageable by coating it and creating non-stick 'liquid marbles', allowing for efficient waste removal

Source: The Biomimicry Institute (2018).

ideas, and innovations. From the table, a few of the numerous amazing functions performed by members of the natural ecosystem are revealed, the combination of which has aided the continuous survival and resilience of organisms in nature.

In this study, resilience is considered a major parameter to be incorporated in the tool for assessing building sustainability. Goals 9 (build resilient infrastructure, promote inclusive and sustainable industrialisation, and foster innovation) and 11 (make cities and human settlements inclusive, safe, resilient, and sustainable) of the United Nations (UN) 2030 SDGs also explicitly corroborate the imperativeness of resilience in realising holistic sustainability. Based on the lessons learned from ES and strategies deployed by natural ecosystems to ensure their survival and resilience in the face of disturbances and disasters (man-made or natural), the following criteria are considered to measure the 'resilience' category in the tool for assessing building sustainability:

a   Disaster management plan.
b   Design considerations for addressing natural disasters.
c   Design considerations for addressing climate change.
d   Use of resilient smart safety technologies; and
e   Optimising landscape elements for resilience.

Gap Two: Inclusivity and Equality

Achieving the goal of sustainability requires a harmonised and balanced incorporation of the three-pronged dimensions (environmental, economic, and social) of sustainability. A major hindrance to achieving this goal is attributed to the significant attention paid to the environmental aspect of sustainability at the expense of the economic and social dimensions. As corroborated by the study of Whang and Kim (2015), there is a wide gap in the number of studies on social and economic issues of sustainability compared to having more than 50% of the global research on sustainable construction (SC) addressing only the environmental dimensions. The social dimension of sustainability has therefore received the least attention with the environmental aspect taking the highest slot, followed keenly by the economic (financial) aspect. According to Segerstedt and Abrahamsson (2019), social sustainability involves desirable, democratic, connected, diverse, and equitable qualities that enhance and result in a good quality of life with a focus on cogent aspects such as the 'walkability' of neighbourhoods, accessibility of facilities and services, quality housing and attractive public realm, mixed tenure, sustainable urban design, social interaction, social networks, active community organisations, thriving cultural traditions, safety, sense of community belonging, social order, well-being, and community cohesion (among and between different groups). Hence, the need to ensure environmental, economic, and social aspects of sustainability are balanced in their integration if the goal of sustainability is to be realised in the construction industry (CI).

More recent practice, discussions, and research on SD have described and identified 'people' as vital decision-makers, innovators, and practitioners and as beneficiaries (Scarlett, 2010). A critical perusal of the UN 2030 SDGs which contains 17 goals, and 169 also emphasises humanity (people) as a core and integral part of the efforts towards achieving holistic sustainability globally. As shown in Table 9.2, Goals 4, 5, 8, 9, 10, 11, and 16 of the SDGs reflect the importance of inclusivity and equality in realising the goal of SD.

The role played by inclusivity and equality in the realisation of an integrated SD cannot be overemphasised. However, achieving the goals of sustainability requires action at all levels and the collective input of the government, agencies,

Table 9.2 Sustainable development goals relating to inclusivity and equality

Goals	Description
SDG4	Ensure inclusive and equitable quality education and promote lifelong learning opportunities for all
SDG5	Achieve gender equality and empower all women and girls
SDG8	Promote sustained, inclusive, and sustainable economic growth, full and productive employment, and decent work for all
SDG9	Build resilient infrastructure, promote inclusive and sustainable industrialisation, and foster innovation
SDG10	Reduce inequality within and among countries
SDG11	Make cities and human settlements inclusive, safe, resilient, and sustainable
SDG16	Promote peaceful and inclusive societies for sustainable development, provide access to justice for all, and build effective, accountable, and inclusive institutions at all levels

Source: UN (2015).

organisations, individuals, and other stakeholders regardless of race, background, ethnicity, gender, and social status, amongst many others. Gender is identified as one of the major continuing elements of economic and social inequalities (Kim et al., 2019). With the equal contribution and effort of everyone (with no exception), there exists a significant potential for the realisation of all the SDGs and other global or country-specific sustainability drives. Within the concept of equality in society, every individual, regardless of their different socio-economic status and abilities, is perceived on the same plane of social relevance (Berg & Schneider, 2012). According to Zabaniotou (2020), gender equality (GE) and women empowerment remain fundamental factors as Goal 5 of the SDGs is critical because its implementation can ensure a positive all-encompassing influence on achieving all other goals and SD. With the issue of gender inequality becoming a widespread phenomenon in all cultures, GE, and women empowerment are now top global agendas in the realisation of holistic sustainability (Bayeh, 2016). As evidenced in the adopted resolutions of the UN for SD, women and girls' empowerment, GE, and inclusiveness are paramount to the prosperity of the planet and human environment (UN, 2015).

Several factors have been identified to impede or lower the incorporation of women in SD which has resulted in the widened gender inequality gap. According to the study of Łapniewska (2019), the low number of women graduates from science, technology, and mathematics (STEM) fields, the perceived technological complexity, and the traditional roles ascribed to women within society are some of the identified factors. Aragonés-González et al. (2020) also identified persisting sexist stereotypes which reinforce gendered cultural roles in learning environments as another hindrance, thereby revealing the importance of coeducation to the promotion of GE. The study of Koralagama et al. (2017) identified three types of gendered discrimination, namely identity-based discrimination, restriction on decision-making, and restriction on access. Ensuring institutional representation, equality, and woman empowerment is therefore vital to achieving inclusive SD.

In the quest for achieving holistic sustainability, Søraa et al. (2020) argued that GE strategies must encompass views that adequately address the forms of intersectionality in strengthening inclusive engagement. The study of Umaña-Barrios and Gil (2017) concluded that significant consideration should be given to women as major stakeholders and users of the public space and amenities in resource optimisation and development towards achieving sustainability goals. As summarised by Mulligan et al. (2018), key attributes of an inclusive built environment highlighted by the Canadian Human Rights Commission (2007) are accessible layout, visible information/directions, and enough size and space to accommodate all persons within a safe environment. The following are a few recommendations from the architectural/design perspective provided by Umaña-Barrios and Gil (2017) and deemed relevant in this study:

a Public telephones and free USB charging ports should be placed close to waiting areas for easy access and in case of emergency;
b Spaces in facilities such as bathrooms (with space for strollers, diaper changing, and baby change facilities) that can be used by both gender parents and their accompanied children dispelling the roles of childcare delegated mainly to women;
c Public seating spaces must be provided to incorporate parents with a free space to place a stroller or a wheelchair right next to it to prevent obstructing free passage;
d The structure and spatial arrangement in public spaces should not obstruct free passage for users, thereby considering disabled persons and respecting the hierarchy of the mobility pyramid (priority for pedestrians, bicycles, and public and private transport);
e Adequate and efficient lighting (using monochrome lights and avoiding those that may affect good visuals) should be provided in spaces to deliver security during day and night hours;
f Maps, timetables, and information routes should be provided with a completely free visual range (lack of these elements can increase the feeling of insecurity among users);
g Cameras for security and systems monitoring need to be installed;
h Spatial accessibility and consideration for users who need support assets (such as parents travelling with strollers, or wheelchairs) to help their mobility is important;
i Multimodality in transportation means and the use of bicycles should be promoted by providing and increasing facilities that encourage their use (since this mode of transportation contributes to lower levels of emissions); and
j There needs to be an efficient arrangement of elements and structures to ensure good visibility thereby increasing users' confidence and visual control.

According to Biswas (2019), equality itself is insufficient to institute an inclusive society despite being regarded as cardinal to achieving inclusive growth. Vital among other key targets such as communication, transparency, commitment, and training, is awareness in the process of transitioning to inclusive management (Sanz-Hernández et al., 2019). Pertinent to achieving the goals of sustainability is, therefore, the stakeholders' adoption of approaches that favour equality and inclusivity (Suckall et al., 2019). According to Whajah et al. (2019), inclusivity aims at providing equal opportunity regardless of income bracket, requiring the growth and improvement in the welfare and societal well-being of individuals. The study of Ros-Tonen et al. (2019) on the inclusive value chain revealed six points constituting the dimensions of inclusiveness, namely double or triple bottom line, concern for well-being, inclusive learning, and innovation, alignment with small-holders' realities, empowerment, gender equity and responsiveness, environmental stability, and enabling environment. The study by Hasan et al. (2019) on disaster management policy in Bangladesh identified gender-based equal participation in decision-making in the context of disaster management; access to early warning system; access to information and resources; gender-based security in shelters; access to relief; gender-based initiative in rescue and rehabilitation; ensuring gender-based income opportunities; identification and assessment of gender-related needs; empowerment of women; and assessment and collection of gender-disaggregated data as core issues for consideration.

Inclusivity in SD can be achieved by the 'simultaneous pursuit of social, environmental and economic development by individuals and/or collectives who, regardless of their socioeconomic status, are empowered to be active and reflexive agents in the identification and prioritisation of threats to global sustainability, as well as the development, planning, and execution of relevant solutions' (Thakhathi, 2019). Inclusive SD or inclusivity incorporates stakeholder engagement approaches. Inclusivity recognises the burden placed on disadvantaged people and communities that prevent them from partaking in the benefits of innovation and development (Woodson et al., 2019). Inclusivity in realising the goals of SD must incorporate all and sundry while ensuring people are treated as stakeholders regardless of their cultural or socioeconomic status. According to Luo et al. (2017), equality of opportunities and favourability of marginalised groups of people, communities, regions, and races are the core considerations in inclusivity. By considering the input of stakeholders, important SD policy issues can be identified, and generationally and culturally sensitive bottom-up policymaking can be influenced (Chapman & Shigetomi, 2018). Also, the inclusive participation of stakeholders is imperative to ensure their commitment, effectiveness, and efficiency (Bijman & Wijers, 2019). Considering the poor social status of people in sub-Saharan Africa, the study of Asongu and Le Roux (2017) suggested the promotion and adoption of information and communication technology (ICT) for more inclusive SD because of its relevance for socioeconomic benefits. As inferred in the study by Luck (2018) on the inclusive design movement, inclusivity ensures redress and change is brought about in all aspects of human life where the different potentialities of people living with disabilities are not considered or accommodated. When the entirety of stakeholders is duly integrated into the decision-making of what directly or indirectly affects them, positive strides would be recorded in achieving inclusive SD. In summary, inclusivity considers all classes of people as fundamental to achieving SD (Sharafutdinov et al., 2019).

As suggested by Sengupta (2016), people must be seen as participants and not as spectators because participation guarantees their confidence, equality, voice, and sense of belonging. Therefore, for inclusiveness to effectively lift low-income communities out of poverty and stimulate rural development, it must be characterised by vital factors such as long-term commitment, adaptation, flexibility, empowerment, capacity building, and trust (Chamberlain & Anseeuw, 2019).

In this study, the consideration of inclusivity and equality as a standalone category for assessing building sustainability is imperative owing to their pivotal role in achieving SD. A perusal of all the SDGs revealed the emphasis placed on ensuring 'all' people are carried along in the process of SD. Also, Goals 4 (ensure inclusive and equitable quality education and promote lifelong learning opportunities for all), 5 (achieve GE and empower all women and girls), 8 (promote sustained, inclusive and sustainable economic growth, full and productive employment and decent work for all), 9 (build resilient infrastructure, promote inclusive and sustainable industrialisation and foster innovation), 10 (reduce inequality within and among countries), 11 (make cities and human settlements inclusive, safe, resilient and sustainable), and 16 (promote peaceful and inclusive societies for SD, provide access to justice for all and build effective, accountable and inclusive institutions at all levels) corroborate the significance of inclusivity and equality in realising SD. To measure the 'inclusivity and equality' category in the tool for assessing building sustainability, the following criteria are considered:

a   Equitable user control;
b   Spatial adaptability for user inclusiveness;
c   Design consideration for cultural, ethnic, and religious differences of occupants; and
d   Consideration for community integration.

Biomimicry Thinking: Inclusivity and Equality in Nature

It is common knowledge that the built environment constitutes a major hindrance and demarcation between the natural and human environment (Schultz, 2002). This barrier has minimised the benefits and extent to which nature can sustainably interface with humans. Other issues such as discrimination along age, gender, disability, sexual orientation, and race lines are rife in human society (Briscoe & Brown, 2020), thereby hindering holistic SD. Also, the display of exclusive ownership and the dichotomy between cooperation and competition among independent groups or individuals fundamentally serve as significant catalysts for deep-rooted conflicts within human society (Rayner, 2010). These issues within the human environment have somehow disrupted and tampered with the sustainable bearing capacity of the natural ecosystem. Hence, embracing diversity, GE, and inclusion is a significant consideration for promoting and measuring ecological restoration and SD (Cid & Bowser, 2015; Ramirez et al., 2018; de Siqueira et al., 2021).

Similar to the human environment, the principle of individuation is operational in the natural world because species in nature exist as distinct entities, each representing an individual within its lineage (Mayden, 2002). Despite their unique-ness and differing characteristics, species in nature have evolved with strategies that support and enhance their integration and cohesion. There are numerous pro-cesses exhibited in the natural ecosystem that show that organisms have embraced inclusivity and equality to ensure their sustenance and sustainability over the years. To optimise their functional structure, natural ecosystems leverage the diversity of species diversity present within them (Allen & Andriani, 2007). In affirmation, researchers suggest that equality and inclusivity as a concept is biologically rooted (Mayr, 2002; Termini & Pang, 2020). Therefore, enhancing methodological pro-gress, comprehensive taxonomic training for evolutionary biologists, and equitable resource allocation will contribute to the heightened effectiveness of integrating nature-based knowledge (Schlick-Steiner et al., 2010).

As suggested by biomimicry proponents, biomimicry thinking perceives the natural world from the 'mentor, model, and measure' perspectives. This is the only way through which sustainability virtues can be inferred from nature in a bit to curb environmental exploitation and degradation (Oguntona & Aigbavboa, 2023). Organisms of the same species, different species, within the same ecosystem and different ecosystems have evolved over the years with remarkable lessons for in-clusivity and equality. For example, the oak tree offers refuge and sustenance to a diverse array of organisms (birds, insects, mammals, lichens, and more), while certain species like squirrels and birds reciprocate by aiding in acorn dispersal, enabling the reproduction and growth of new oak trees (The Biomimicry Institute, 2021). Also when feeding with conspecifics, pelicans often exhibit cooperative be-haviour known as 'cooperative herding', where they harness the individual capac-ity of fellow pelicans to drive fish into shallow water or surround them in more open spaces, resulting in a higher catch rate for groups of two to six birds compared to single birds or larger groups (Anderson, 1991). Moreover, their attraction to specific areas was confirmed through tests with decoys that revealed their affinity towards locations occupied by other pelicans.

Another example is the mutual interactions between the sea anemone and clown-fish as obligatory symbionts thereby enabling them to coevolve. As revealed by Lubbock (1980), the clownfish species, *Amphiprion clarkii* exhibits an intriguing ability to coexist unharmed within the tentacles of the sea anemone *Stichodactyla haddoni*, which possesses potent stinging capabilities that would typically capture any non symbiotic fish that ventured into its territory. However, the presence of clownfish influences the behaviour of the anemone without compromising its sting-ing ability, suggesting the absence of a general inhibitory effect mediated by the anemone's nervous system. This clownfish achieves protection from stinging by uti-lising its external mucus layer, which is notably thicker (around three to four times) compared to related fish species that do not inhabit anemones and are primarily composed of glycoprotein-containing neutral polysaccharides. Also, the early warn-ing system (alerts) provided by the black-capped Chickadee enhances the survival rate of multiple species within and around a particular ecosystem. As indicated by Templeton et al. (2005), this remarkable system, characterised as one of the most so-phisticated and nuanced signalling systems ever observed, stands out for its unique combination of referential and risk-based antipredator vocalisation components. Chickadees employ a 'seet' alarm call to signal the presence of a swiftly moving predator, such as a flying raptor, while a distinct 'chick-a-dee' mobbing call is utilised when confronting a stationary predator like a perched raptor. This suggests that these two vocalisations serve as functional references to specific predator encounters, enabling chickadees to convey information about predators at multiple levels. The array of processes and strategies domiciled and exhibited in the natural world strongly conveys the message of inclusivity and equality which has enabled their evolution and survival over the years. Hence, the conscious adoption and implementation of inclusivity and equality cannot be overemphasised if SD is to be achieved.

Gap Three: Collaboration

Achieving According to The Land Institute (2020), 'when people, land, and community are as one, all three members prosper; when they relate not as members but as competing interests, all three are exploited...'. Therefore, collaboration is significant and the only avenue to create new systems, methodologies, and approaches to optimise growth and efficiency in ways that address the global challenges facing our already resource-constrained world (Böggemann & Both, 2014). Global collaboration among stakeholders on an exceptional level is pivotal and required to achieve SD (Fiksel, 2009). The collaborative partnership has also been identified as significant to the implementation of the UN 2030 SDGs with Goal 17 (strengthen the means of implementation and revitalise the global partnership of SD) singly dedicated to this effect. Also, the medium to the long-term future sustainability of cities will be influenced based on infrastructural decisions made today (Kenny et al., 2012). As further noted by the United Nations Environment Programme (UNEP) publication, the decisions and choices made on the future sustainability of infrastructures should encourage resilient global and local collaboration in human societies (Peter & Swilling, 2012). To successfully implement and achieve the SDGs and SD, the study of Kourula et al. (2017) suggested the development of collaborative networks and partnerships within communities, organisations (intergovernmental and nongovernmental), government, and other relevant stakeholders.

Directly engaging local communities is critical to achieving the SDGs and SD. Owing to their attachment to the community (as they identify with and draw their livelihoods from the community), local indigenous people possess outstanding knowledge of the situation, experience, and time of the community. According to Scarlett (2010), collaborating with the locals offers local knowledge which reflects local circumstances, cultures, and priorities in maximising economic, social, and environmental benefits. By engaging with the local communities and integrating them into the decision-making process as stakeholders, they could offer innovative insights based on their in-depth knowledge possessed over the years on contexts and practices affecting the sustainability of their social and natural systems (Power & Chapin, 2010). The knowledge (raw and refined) offered by the locals also can protect the environmental ecosystem and enhance the socio-economic status of the general populace in the process of achieving SD.

To drive sustainability at speed and scale, collaboration (skill and will to partner with stakeholders) is a key factor (Grayson, 2019). Effective collaboration is dependent on a two-way link based on attributes such as shared power and non-hierarchical relationships, shared planning and decision-making, and cooperation (Gassner et al., 1999). The study of Yang and Lin (2020) on supply chain affirmed that collaboration is an effective strategy for economic and environmental protection and performance improvement. As indicated by the study of Luo et al. (2010), the Theory of Remote Scientific Collaboration (TORSC) asserts five critical success factors of collaboration, namely the nature of the work (tasks involved in distributed collaborative work that range from highly interdependent to easily modularised work); management planning, and decision making (having frequent real-time coordination and communication); common ground (having mutual knowledge, beliefs, and assumptions regarding management style); collaboration readiness (collaborators have been motivated to collaborate, enjoy working together and trusting each other); and technology readiness (participants having the appropriate technology to complete their collaborative tasks and comfortable using that technology). In the glocal model, engagement for local learning and impact, experiences, and the utilisation of digital technologies are combined in transnational collaboration to achieve SD in an increasingly interconnected and globalised world (Caniglia et al., 2018). As summarised by the study of Morel et al. (2020), the collaborative governance model identified leadership, consequential incentives, interdependence between public and private actors, and uncertainty about the problem and its possible solutions as driving factors. Valuing the needs of all stakeholders, building trust, frank discussion of concerns of all stakeholders, cumulative advantages for members, similar attributes and interests, shared focus, and empowering community stakeholders are identified by the study of Green et al. (2016) as key elements of successful collaboration. The study of Feng et al. (2020) on green supplier collaboration identified trust and dependence as key factors that influence the effectiveness of collaboration. Four factors, namely drivers (refers to socio-political factors that promote the formation of partnerships to achieve sustainability objectives); motivations (refers to the understanding of different intentions of partners as these differences can result in a mismatch within the partnership); partner and partnership characteristics (refers to the cogent attributes of partner and partnerships, which are imperative for partner selection and partnership design); and process (refers to the nature of interactions among the partners in their engagements towards achieving the goals) are identified by Gray and Stites (2013) as critical to the success of the collaboration.

Simon Zadek, a visiting fellow at the Tsinghua School of Economics and Management in Beijing, said, 'Partners think that collaboration will change the world. Then it doesn't, and they think that it failed. But often the collaboration changed something – the way some part of the system works and delivers outcomes. It is a matter of understanding the nature of change itself'. Therefore, it is important to remember that while collaboration may not result in the intended outcomes, it has the potential to positively influence the processes and strategies (Albani & Henderson, 2014). A study by Chen et al. (2017) on supply chain collaboration for sustainability discovered that 70% to 80% of the literature focuses on collaboration concerning the environmental and economic aspects while 20% to 30% only addresses social collaboration.

Numerous collaborative efforts to tackle global challenges (such as ecosystem loss, resource depletion, and climate change) have failed to yield desired results owing to factors such as shortage of trust, lack of shared purpose, and competitive self-interest (Nidumolu et al., 2014). As identified by Vazquez-Brust et al. (2020), other issues affecting the success of collaboration in positively influencing the achievement of SD include inappropriate collaborative governance structures, poor leadership and inadequate fit to problem structure, discontinued funding, and wrong partner mix. The study of Wondirad et al. (2020) on ecotourism identified ten factors hindering stakeholder collaboration: poor governance; lack of awareness amongst stakeholders about the relevance of collaboration; poor societal culture of collaboration; resource constraints; lack of trust and mutual understanding amongst stakeholders; lack of sufficient and sustained communication amongst stakeholders; the limited size of a sector in the country and region receiving little attention; the existence of diverse interests and unhealthy competition amongst stakeholders; power friction within governmental organisations and amongst government, local communities, and private enterprises; and conflicts amongst ethnic tribes. Conflict is another detrimental factor that hampers the effect of collaboration in achieving SD (Curşeu & Schruijer, 2017). Other barriers to the success of collaboration are lack of trust, organisational boundaries, conflicting objectives, and values, lack of awareness, differing social norms, and lack of knowledge (Morel et al., 2020). By overcoming all these hindrances, the collaboration will be able to enhance economic, social, and environmental performance and offer improved benefits to facilitate a smooth transition to SD (Niesten et al., 2017).

Biomimicry Thinking: Collaboration in Nature

According to Camarinha-Matos and Afsarmanesh (2018), the concept of collaboration has its roots in nature (the beginning of life on earth) and was not originally started by any of the sectors (e.g., management, social sciences, computer sciences, industrial engineering) that address it today. Outstanding collaborative efforts and processes are identified by studying the natural ecosystem. For plants, animals, and other organisms to survive for over 3.8 billion years of their evolution, numerous collaborations are identified to be responsible, as exhibited by the smallest to the biggest organisms in nature. With the emerging field of biomimicry, the understanding of the lessons in nature is studied and emulated to solve human challenges sustainably. According to Kellert (2014), biomimicry examines amongst many others a leopard's vision, a bat's echolocation, a snake's venom, a shark's skin, and a spider's web and infers a wealth of solutions for optimising the sustainability of the human environment. Table 9.3 shows a few examples of nature's strategies for collaboration. According to Angne (2012), it is imperative for biologists and other stakeholders with robust knowledge and understanding

Table 9.3 Nature examples of collaboration

S/N	Function	Nature inspiration	Strategy
1	Cooperate within the same species, coordinate by self-organisation	Ants, honey bees	Honeybees collaborate (for group decisions) when foraging, selecting a new hive through knowledge sharing, and ants' behaviour when they are under threat
2	Cooperate within an ecosystem, cooperate/ compete between different species, coordinate activities, cycle nutrients	Mangroves	Several species of epiphytes, ants, fungi, and butterflies in mangrove forests provide benefits to each other through mutualism
3	Capture, absorb, or filter chemical entities, cooperate/compete between different species	Tree Bugs, cicadas, leafhoppers, aphids, etc.	The nutrient-poor diet of cicadas is supplemented thanks to specialised bacterial symbionts
4	Differentiate signal from noise, maintain homeostasis, respond to signals	Human	Neurons aid organisms in reacting to environmental stimuli because they collaborate to sense the environment, share information, and filter unimportant information
5	Protect from light	Common ivy	Organic nanoparticles secreted by English ivy rootlets absorb and scatter ultraviolet light thanks to large surface-to-volume ratio and uniformity

Source: The Biomimicry Institute (2018).

of the natural ecosystem to be at the design table as their input will be required depending on the complexity of the issues at hand while bearing in mind that biomimicry employs collaboration to achieve its overarching goal of sustainability. By ensuring collaboration among professionals and other stakeholders across disciplines and sectors, solutions will be proffered to lower the impact of the numerous environmental challenges facing humanity (Garcia, 2017). In the spirit of true accommodation where differences are respected, stakeholders are significant contributors, no party dominates, and compromise is seen as neither defeat nor conflict which remain vital factors for a successful collaboration (Kellert, 2014) that will see biomimicry result in sustainable solutions. Numerous examples of collaboration processes are found to be abundant in nature and reflected in the variety of attributes of the involved organisms. With a strong emphasis on a robust collaboration that cuts across all disciplines involving experts and other stakeholders within the value chain, biomimicry implementation will result in successful outcomes (Tempelman et al., 2015).

This study considers collaboration as a gap, as inferred from the review of the existing GBRTs and SATs, owing to its significance in achieving SD. Goal 17 (strengthen the means of implementation and revitalise the global partnership of

SD) of the SDGs which points to collaboration among 'everyone' is also noted as pivotal to the achievement of other goals. To measure the 'collaboration' category in the tool for assessing building sustainability, the following criteria are considered:

a Effective multidisciplinary collaboration;
b Effective stakeholders conflict management;
c Effective collaboration with locals; and
d Optimised corporate social responsibility.

Summary

This chapter discussed the gaps identified in the studies on SATs in a bid to satisfy the objective of this book. The trio of inclusivity and equality, collaboration, and resilience are identified and presented based on a comprehensive review of existing SATs and GBRTs that are utilised globally. To achieve a nature-inspired SAT that holistically addresses the three pillars of sustainability, these gaps are found to be significant. Lastly, this chapter discusses the root of the identified gaps in nature and presents a few examples of how the natural ecosystem has exhibited these cogent concepts. The next chapter presents the framework of the conceptualised Biomimicy Sustainability Assessment Tool (BioSAT) for the CI. The chapter also presents the constructs that form the bedrock of the conceptualised assessment tool.

References

Achour, N., Pantzartzis, E., Pascale, F., & Price, A. D. (2015). Integration of resilience and sustainability: From theory to application. *International Journal of Disaster Resilience in the Built Environment, 6*(3), 347–362.

Adom, D., Hussein, E. K., & Agyem, J. A. (2018). Theoretical and conceptual framework: Mandatory ingredients of a quality research. *International Journal of Scientific Research, 7*(1), 438–441.

Albani, M., & Henderson, K. (2014). *Creating partnerships for sustainability.* Retrieved 21 January 2020 from https://www.mckinsey.com/business-functions/sustainability/our-insights/creating-partnerships-for-sustainability

Allam, Z. (2020). The triple B: Big data, biotechnology, and biomimicry. *Biotechnology and future cities* (pp. 17–33) Palgrave Macmillan, Cham.

Allen, P. M., & Andriani, P. (2007). Diversity, interconnectivity and sustainability. In *Complexity, science and society* (pp. 11–32). CRC Press.

Anderies, J. M., Folke, C., Walker, B., & Ostrom, E. (2013). Aligning key concepts for global change policy: Robustness, resilience, and sustainability. *Ecology and Society, 18*(2), 1–16.

Anderson, J. G. (1991). Foraging behavior of the American white pelican (Pelecanus erythrorhyncos) in Western Nevada. *Colonial Waterbirds, 2*, 166–172.

Angne, S. M. (2012). Biomimicry: An interior design teaching tool. In Proceedings of Second Annual Biomimicry in Higher Education. Webinar, online. Retrieved 21 January 2018 from https://b38website.azurewebsites.net/resource-categories/shop-biomimicry/

Aragonés-González, M., Rosser-Limiñana, A., & Gil-González, D. (2020). Coeducation and gender equality in education systems: A scoping review. *Children and Youth Services Review, 111*, 1–12.

Asongu, S. A., & Le Roux, S. (2017). Enhancing ICT for inclusive human development in sub-Saharan Africa. *Technological Forecasting & Social Change, 118*, 44–54.

Bayeh, E. (2016). The role of empowering women and achieving gender equality to the sustainable development of Ethiopia. *Pacific Science Review B: Humanities and Social Sciences, 2*(1), 37–42.

Benyus, J. M. (2011). *A biomimicry primer* Biomimicry 3.8.

Berg, D. H., & Schneider, C. (2012). Equality dichotomies in inclusive education: Comparing Canada and France. *Alter – European Journal of Disability Research, Revue Européen De Recherche Sur Le Handicap, 6*(2), 124–134.

Bijman, J., & Wijers, G. (2019). Exploring the inclusiveness of producer cooperatives. *Current Opinion in Environmental Sustainability, 41*, 74–79.

Biswas, A. (2019). A framework to analyse inclusiveness of urban policy. *Cities, 87*, 174–184.

Böggemann, M., & Both, N. (2014). The resilience action initiative: An introduction. In R. Kupers (Ed.), *Turbulence* (pp. 23–34). Netherlands: Amsterdam University Press.

Bousquet, F., Botta, A., Alinovi, L., Barreteau, O., Bossio, D., Brown, K., Caron, P., Cury, P., d'Errico, M., DeClerck, F., Dessard, H., Kautsky, E. E., Fabricius, C., Folke, C., Fortmann, L., Hubert, B., Magda, D., Mathevet, R., Norgaard, R. B., Quinlan, A., & Staver, C. (2016). Resilience and development: Mobilizing for transformation. *Ecology and Society, 21*(3), 1–18.

Briscoe, J., & Brown, K. (2020). Inclusion and diversity in developmental biology: Introducing the node network. *Development, 147*(2), dev187591.

Camarinha-Matos, L. M., & Afsarmanesh, H. (2018). Roots of collaboration: Nature-inspired solutions for collaborative networks. *IEEE Access, 6*, 30829–30843.

Canadian Human Rights Commission (2007). *International best practices in universal design: A global review*. Ottawa, Canada: Betty Dion Enterprises.

Caniglia, G., John, B., Bellina, L., Lang, D. J., Wiek, A., Cohmer, S., & Laubichler, M. D. (2018). The glocal curriculum: A model for transnational collaboration in higher education for sustainable development. *Journal of Cleaner Production, 171*, 368–376.

Carlson, J. L., Haffenden, R. A., Bassett, G. W., Buehring, W. A., Collins, I. M. J., & Folga, S. M., et al. (2012). *Resilience: Theory and application*. U.S Department of Energy, USA.

Chamberlain, W., & Anseeuw, W. (2019). Inclusive businesses in agriculture: Defining the concept and its complex and evolving partnership structures in the field. *Land Use Policy, 83*, 308–322.

Chan, K. M. A., Shaw, M. R., Cameron, D. R., Underwood, E. C., & Daily, G. C. (2006). Conservation planning for ecosystem services. *PLoS Biology, 4*(11), 2138–2152.

Chapman, A., & Shigetomi, Y. (2018). Developing national frameworks for inclusive sustainable development incorporating lifestyle factor importance. *Journal of Cleaner Production, 200*, 39–47.

Chen, L., Zhao, X., Tang, O., Price, L., Zhang, S., & Zhu, W. (2017). Supply chain collaboration for sustainability: A literature review and future research agenda. *International Journal of Production Economics, 194*, 73–87.

Cid, C. R., & Bowser, G. (2015). Breaking down the barriers to diversity in ecology. *Frontiers in Ecology and the Environment, 13*(4), 179–179.

Curşeu, P. L., & Schruijer, S. G. (2017). Stakeholder diversity and the comprehensiveness of sustainability decisions: The role of collaboration and conflict. *Current Opinion in Environmental Sustainability, 28*, 114–120.

Daily, G. C. (2003). What are ecosystem services? In Lorey, D. E. (Ed.), *Global environmental challenges of the twenty-first century: Resources, consumption, and sustainable solutions* (pp. 227–232). USA: Scholarly Resources Inc.

de Siqueira, L. P., Tedesco, A. M., Meli, P., Diederichsen, A., & Brancalion, P. H. (2021). Gender inclusion in ecological restoration. *Restoration Ecology, 29*(7), e13497.

Desjardins, E., Barker, G., Lindo, Z., Dieleman, C., & Dussault, A. C. (2015). Promoting resilience. *The Quarterly Review of Biology, 90*(2), 147–165.

Elmqvist, T., Setälä, H., Handel, S. N., Van Der Ploeg, S., Aronson, J., Blignaut, J. N., Gomez-Baggethun, E., Nowak, D. J., Kronenberg, J., & De Groot, R. (2015). Benefits of restoring ecosystem services in urban areas. *Current Opinion in Environmental Sustainability, 14*, 101–108.

Evans, B., & Reid, J. (2013). Dangerously exposed: The life and death of the resilient subject. *Resilience, 1*(2), 83–98.

Feng, T., Jiang, Y., & Xu, D. (2020). The dual-process between green supplier collaboration and firm performance: A behavioral perspective. *Journal of Cleaner Production, 260*, 1–17.

Fiksel, J. (2009). *Design for environment: A guide to sustainable product development* (2nd ed.). USA: The McGraw-Hill.

Fisher, B., Turner, R. K., & Morling, P. (2009). Defining and classifying ecosystem services for decision making. *Ecological Economics, 68*(3), 643–653.

Folke, C., Jansson, Å, Rockström, J., Olsson, P., Carpenter, S. R., Chapin, F. S., Crépin, A. S., Daily, G., Danell, K., Ebbesson, J., Elmqvist, T., Galaz, V., Moberg, F., Nilsson, M., Österblom, H., Ostrom, E., Persson, Å, Peterson, G., Polasky, S., Steffen, W., Walker, B., & Westley, F. (2011). Reconnecting to the biosphere. *Ambio, 40*(7), 719–738.

Folke, C., Carpenter, S. R., Walker, B., Scheffer, M., Chapin, T., & Rockström, J. (2010). Resilience thinking; Integrating resilience, adaptability, and transformability. *Ecology and Society, 15*(4), 1–9.

Galaz, V. (2005). Social-ecological resilience And social conflict: Institutions And strategic adaptation in Swedish water management. *Ambio, 34*(7), 567–572.

Garcia, P. R. (2017). The influence of the concepts of biophilia and biomimicry in contemporary architecture. *Journal of Civil Engineering and Architecture, 11*(5), 500–513.

Gassner, L., Wotton, K., Clare, J., Hofmeyer, A., & Buckman, J. (1999). Theory meets practice: Evaluation of a model of collaboration: Academic and clinician partnership in the development and implementation teaching. *Collegian, 6*(3), 14–28.

Gray, B., & Stites, J. P. (2013). *Sustainability through partnerships: Capitalizing on collaboration*. Ontario, Canada: Network for Business Sustainability.

Grayson, D. (2019). *Collaborate for success and sustainability.* Retrieved 22 January 2020 from https://www.globalfocusmagazine.com/collaborate-for-success-and-sustainability/

Green, A. E., Trott, E., Willging, C. E., Finn, N. K., Ehrhart, M. G., & Aarons, G. A. (2016). The role of collaborations in sustaining an evidence-based intervention to reduce child neglect. *Child Abuse & Neglect, 53*, 4–16.

Habib, M. K., & Nagata, F. (2019). *Bioinspired design: Creativity and sustainability*. In Proceedings of 20th International Conference on Research and Education in Mechatronics (REM), Wels, Austria, 1–5.

Hargroves, K., & Smith, M. (2006). Innovation inspired by nature: Biomimicry. *Ecos, 2006*(129), 27–29.

Hasan, M. R., Nasreen, M., & Chowdhury, M. A. (2019). Gender-inclusive disaster management policy in Bangladesh: A content analysis of national and international regulatory frameworks. *International Journal of Disaster Risk Reduction, 41*, 1–10.

Holling, C. S. (1973). Resilience and stability of ecological systems. *Annual Review of Ecology and Systematics, 4*(1), 1–23.

Hydes, K., Richardson, G. R., & Petinelli, G. (2018). World green building council: Supporting the sustainable transformation of the global property market. *World Green Building Council: Supporting the Sustainable Transformation of the Global Property Market,* 49–52.

Kellert, S. (2014). Biophilia and biomimicry: Evolutionary adaptation of human versus nonhuman nature. *Intelligent Buildings International, 8*(2), 1–6.

Kenny, J., Desha, C., Kumar, A., & Hargroves, C. (2012). Using biomimicry to inform urban infrastructure design that addresses 21st-century needs. *1st International Conference on Urban Sustainability and Resilience: Conference Proceedings,* 1–13.

Kim, K., İlkkaracan, İ, & Kaya, T. (2019). Public investment in care services in Turkey: Promoting employment & gender inclusive growth. *Journal of Policy Modeling, 41*(6), 1210–1229.

Koralagama, D., Gupta, J., & Pouw, N. (2017). Inclusive development from a gender perspective in small-scale fisheries. *Current Opinion in Environmental Sustainability, 24,* 1–6.

Kourula, A., Pisani, N., & Kolk, A. (2017). Corporate sustainability and inclusive development: Highlights from international business and management research. *Current Opinion in Environmental Sustainability, 24,* 14–18.

Kunz, T. H., Braun de Torrez, E., Bauer, D., Lobova, T., & Fleming, T. H. (2011). Ecosystem services provided by bats. *Annals of the New York Academy of Sciences, 1223*(1), 1–38.

Langeland, K. S., Manheim, D., McLeod, G., & Nacouzi, G. (2016). Definitions, characteristics, and assessments of resilience. In *How civil institutions build resilience* (pp. 5–10). Santa Monica, CA: RAND Corporation.

Łapniewska, Z. (2019). Energy, equality and sustainability? European Electricity cooperatives from a gender perspective. *Energy Research & Social Science, 57,* 1–12.

Ledesma, J. (2014). Conceptual frameworks and research models on resilience in leadership. *SAGE Open, 4*(3), 1–8.

Liu, A., Teo, I., Chen, D., Lu, S., Wuest, T., Zhang, Z., & Tao, F. (2019). Biologically inspired design of context-aware smart products. *Engineering, 5*(4), 637–645.

Lubbock, R. (1980). Why are clownfishes not stung by sea anemones? *Proceedings of the Royal Society of London. Series B. Biological Sciences, 207*(1166), 35–61.

Luck, R. (2018). Inclusive design and making in practice: Bringing bodily experience into closer contact with making. *Design Studies, 54,* 96–119.

Luo, X., Muleta, D., Hu, Z., Tang, H., Zhao, Z., Shen, S., & Lee, B. (2017). Inclusive development and agricultural adaptation to climate change. *Current Opinion in Environmental Sustainability, 24,* 78–83.

Luo, A., Ng'ambi, D., & Hanss, T. (2010). Towards building a productive, scalable and sustainable collaboration model for open educational resources. *In Proceedings of the 16th ACM international conference on Supporting group work,* 273–282.

Marchese, D., Reynolds, E., Bates, M. E., Morgan, H., Clark, S. S., & Linkov, I. (2018). Resilience and sustainability: Similarities and differences in environmental management applications. *Science of the Total Environment, 613-614,* 1275–1283.

Markelj, J., Kitek Kuzman, M., Grošelj, P., & Zbašnik-Senegačnik, M. (2014). A simplified method for evaluating building sustainability in the early design phase for architects. *Sustainability, 6*(12), 8775–8795.

Mayden, R. L. (2002). On biological species, species concepts and individuation in the natural world. *Fish and Fisheries, 3*(3), 171–196.

Mayr, E. (2002). The biology of race and the concept of equality. *Daedalus, 131*(1), 89–94.

McPhearson, T. (2014). *The rise of resilience: Linking resilience and sustainability in city planning*. Retrieved 20 November 2018 from https://www.thenatureofcities.com/2014/06/08/the-rise-of-resilience-linking-resilience-and-sustainability-in-city-planning/

Millennium Ecosystem Assessment (2005). *Ecosystems and human well-being: Synthesis no. 5*. Washington, DC: Island Press.

Morel, M., Balm, S., Berden, M., & van Amstel, W. P. (2020). Governance models for sustainable urban construction logistics: Barriers for collaboration. *Transportation Research Procedia, 46*, 173–180.

Mulligan, K., Calder, A., & Mulligan, H. (2018). Inclusive design in architectural practice: Experiential learning of disability in architectural education. *Disability and Health Journal, 11*(2), 237–242.

Nidumolu, R., Ellison, J., Whalen, J., & Billman, E. (2014). *The collaboration imperative. Harvard Business Review, 92*(4), 76–84.

Niesten, E., Jolink, A., de Sousa Jabbour, A. B. L., Chappin, M., & Lozano, R. (2017). Sustainable collaboration: The impact of governance and institutions on sustainable performance. *Journal of Cleaner Production, 155*(Part 2), 1–6.

Oguntona, O. A., & Aigbavboa, C. O. (2023). Nature inspiration, imitation, and emulation: Biomimicry thinking path to sustainability in the construction industry. *Frontiers in Built Environment, 9*, 1085979.

Olsson, P., Galaz, V., & Boonstra, W. J. (2014). Sustainability transformations: A resilience perspective. *Ecology and Society, 19*(4), 1–13.

Peter, C., & Swilling, M. (2012). *Sustainable, resource efficient cities: Making it happen!* France: United Nations Environment Programme.

Power, M. E., & Chapin, F. S. III (2010). Planetary stewardship in a changing world: Paths towards resilience and sustainability. *Bulletin of the Ecological Society of America, 91*(2), 143–175.

Rademaker, M., Jans, K., Verhagen, P., Boeschoten, A., Roos, H., & Slingerland, S. (2018). *Resilience from national to local level*. Netherlands: Hague Centre for Strategic Studies.

Ramirez, K. S., Berhe, A. A., Burt, J., Gil-Romera, G., Johnson, R. F., Koltz, A. M., & Tuff, K. (2018). The future of ecology is collaborative, inclusive and deconstructs biases. *Nature Ecology & Evolution, 2*(2), 200–200.

Rayner, A. D. (2010). Inclusionality and sustainability – Attuning with the currency of natural energy flow and how this contrasts with abstract economic rationality. *Environmental Economics, 1*(1), 102–112.

Redman, C. L. (2014). Should sustainability and resilience be combined or remain distinct pursuits? *Ecology and Society, 19*(2), 1–8.

Rogers, P. (2015). Researching resilience: An agenda for change. *Resilience, 3*(1), 55–71.

Romero-Lankao, P., Gnatz, D., Wilhelmi, O., & Hayden, M. (2016). Urban sustainability and resilience: From theory to practice. *Sustainability, 8*(1224), 1–19.

Ros-Tonen, M. A., Bitzer, V., Laven, A., Ollivier de Leth, D., Van Leynseele, Y., & Vos, A. (2019). Conceptualizing inclusiveness of smallholder value chain integration. *Current Opinion in Environmental Sustainability, 41*, 10–17.

Saito, O., & Ryu, H. (2020). What and how are we sharing? The academic landscape of The sharing paradigm and practices: Objectives and organization of The book. In O. Saito (Ed.), *Sharing ecosystem services* (pp. 1–20). Singapore: Springer.

Sanz-Hernández, A., Sanagustín-Fons, M. V., & López-Rodríguez, M. E. (2019). A transition to an innovative and inclusive bioeconomy in Aragon, Spain. *Environmental Innovation and Societal Transitions, 33*, 301–316.

Say, C., & Wood, A. (2008). Sustainable rating systems around the world. *Council on Tall Buildings and Urban Habitat Journal (CTBUH Review)*, *2*, 18–29.

Scarlett, L. (2010). Cities and sustainability-ecology, economy, and community. *Sustainable Development Law & Policy*, *XI*(1), 1–74.

Schlick-Steiner, B. C., Steiner, F. M., Seifert, B., Stauffer, C., Christian, E., & Crozier, R. H. (2010). Integrative taxonomy: A multisource approach to exploring biodiversity. *Annual Review of Entomology*, *55*, 421–438.

Schultz, P. W. (2002). Inclusion with nature: The psychology of human-nature relations. In *Psychology of sustainable development* (pp. 61–78). Boston, MA: Springer US.

Scott, L. (2017). *Lessons in resilience: From biological systems to human food systems*. US: Oberlin: Oberlin College and Conservatory.

Segerstedt, E., & Abrahamsson, L. (2019). Diversity of livelihoods and social sustainability in established mining communities. *The Extractive Industries and Society*, *6*(2), 610–619.

Sengupta, P. (2016). How effective is inclusive innovation without participation? *Geoforum*, *75*, 12–15.

Sharafutdinov, R. I., Akhmetshin, E. M., Polyakova, A. G., Gerasimov, V. O., Shpakova, R. N., & Mikhailova, M. V. (2019). Inclusive growth: A dataset on key and institutional foundations for inclusive development of Russian regions. *Data in Brief*, *23*, 1–11.

Shean, M. B. (2015). *Current theories relating to resilience and young people: A literature review*. Melbourne: VicHealth.

Søraa, R. A., Anfinsen, M., Foulds, C., Korsnes, M., Lagesen, V., Robison, R., & Ryghaug, M. (2020). Diversifying diversity: Inclusive engagement, intersectionality, and gender identity in a European social sciences and humanities energy research project. *Energy Research & Social Science*, *62*, 1–11.

Stevens, B., & Schieb, P. (2007). Infrastructure to 2030: Main findings and policy recommendations. In Organisation for Economic Co-Operation and Development, (OECD) (Ed.), *Infrastructure to 2030: Mapping policy for electricity, water and transport*, *2*, 19–106. Paris: OECD Publishing.

Suckall, N., Tompkins, E. L., & Vincent, K. (2019). A framework to analyse the implications of coastal transformation on inclusive development. *Environmental Science and Policy*, *96*, 64–69.

Tempelman, E., Pauw, I. C. D., Grinten, B. V. D., Mul, E. J., & Grevers, K. (2015). Biomimicry and cradle to cradle in product design: An analysis of current design practice. *Journal of Design Research*, *13*(4), 326–344.

Templeton, C. N., Greene, E., & Davis, K. (2005). Allometry of alarm calls: Black-capped chickadees encode information about predator size. *Science*, *308*(5730), 1934–1937.

Termini, C. M., & Pang, A. (2020). Beyond the bench: How inclusion and exclusion make us the scientists we are. *Molecular Biology of the Cell*, *31*(20), 2164–2167.

Thakhathi, A. (2019). Creative start-up capital raising for inclusive sustainable development: A case study of Boswa ba rona development corporation's self-reliance. *Journal of Cleaner Production*, *241*, 1–12.

The Biomimicry Institute (2021). *Asknature*. Retrieved 21 April 2023 from https://asknature.org/?s=&page=0&hFR%5Bpost_type_label%5D%5B0%5D=Biological%20Strategies&hFR%5Btaxonomies_hierarchical.function.lvl0%5D%5B0%5D=Maintain%20Community&is_v=1

The Land Institute. (2020). *Vision & mission*. Retrieved 15 February 2020 from https://landinstitute.org/about-us/vision-mission/

Thiede, B. (2016). Resilience and development among ultra-poor households in rural Ethiopia. *Resilience*, *4*(1), 1–13.

Turner, B. L. (2010). Vulnerability and resilience: Coalescing or paralleling approaches for sustainability science? *Global Environmental Change, 20*(4), 570–576.

Umaña-Barrios, N., & Gil, A. S. (2017). How can spatial design promote inclusivity, gender equality and overall sustainability in Costa Rica's urban mobility system? *Procedia Engineering, 198,* 1018–1035.

United Nations. (2015). *Transforming our world: The 2030 Agenda for Sustainable Development.* Retrieved 24 September 2019 from https://sustainabledevelopment.un.org/post2015/transformingourworld

Van der Heijden, J. (2017). Urban sustainability and resilience. In P. Drahos (Ed.), *Regulatory theory* (pp. 725–740). Australia: Australian National University (ANU) Press.

Van Wyk, L. (2015). From sustainability to resilience: A paradigm shift. In L. van Wyk (Ed.), *Green building handbook* (pp. 26–39). South Africa: Alive2Green.

Vazquez-Brust, D., Piao, R. S., de Melo, M. F. D. S., Yaryd, R. T., & Carvalho, M. M. (2020). The governance of collaboration for sustainable development: Exploring the black box. *Journal of Cleaner Production, 256,* 1–12.

Vrasti, W., & Michelsen, N. (2017). Introduction: On resilience and solidarity. *Resilience: Solidarity and Resilience, 5*(1), 1–9.

Walker, B., Kinzig, A., & Langridge, J. (1999). Original articles: Plant attribute diversity, resilience, and ecosystem function: The nature and significance of dominant and minor species. *Ecosystems, 2*(2), 95–113.

Walsh-Dilley, M., & Wolford, W. (2015). (Un)defining resilience: Subjective understandings of 'resilience' from the field. *Resilience: Boundary Objects, Border Thinking: Subjective Understandings of Resilience from the Field, 3*(3), 173–182.

Whajah, J., Bokpin, G. A., & Kuttu, S. (2019). Government size, public debt and inclusive growth in Africa. *Research in International Business and Finance, 49,* 225–240.

Whang, S., & Kim, S. (2015). Balanced sustainable implementation in the construction industry: The perspective of Korean contractors. *Energy and Buildings, 96,* 76–85.

Wondirad, A., Tolkach, D., & King, B. (2020). Stakeholder collaboration as a major factor for sustainable ecotourism development in developing countries. *Tourism Management, 78,* 1–21.

Woodson, T., Alcantara, J. T., & do Nascimento, M. S. (2019). Is 3D printing an inclusive innovation? An examination of 3D printing in Brazil. *Technovation, 80-81,* 54–62.

World Green Building Council. (2020). Green building council South Africa. Retrieved 24 March 2020 from https://www.worldgbc.org/member-directory/green-building-council-south-africa

Yang, Z., & Lin, Y. (2020). The effects of supply chain collaboration on green innovation performance: An interpretive structural modeling analysis. *Sustainable Production and Consumption, 23,* 1–10.

Zabaniotou, A. (2020). Towards gender equality in Mediterranean engineering schools through networking, collaborative learning, synergies and commitment to SDGs – the RMEI approach. *Global Transitions, 2,* 4–15.

Zimmerman, M. A. (2013). Resiliency theory: A strengths-based approach to research and practice for adolescent health. *Health Education & Behavior, 40*(4), 381–383.

10 The Conceptual Biomimicry Sustainability Assessment Tool

Introduction

In the quest to identify and assemble the necessary categories and criteria/indicators for biomimicry sustainability assessment tool (BioSAT) for the construction industry (CI), a critical and strategic evaluation of existing green building rating tools (GBRTs) and sustainability assessment tools (SATs) was conducted. This task was carried out earlier in this book, presenting the constituting evaluation categories, criteria/indicators, and sub-criteria of the reviewed GBRTs and SATs. However, the purpose of the review was not to challenge or contradict the rationality of specific attributes of sustainable development (SD), but rather it aims to identify and collate the influential attributes necessary to develop a comprehensive SAT for assessing the sustainability performance of building projects.

The review of the literature was centred on SD concepts, tools, frameworks, and guidelines that aid the evaluation and determination of how green a building is. Based on the understanding of what constitutes SD in its comprehensive state, the categories, criteria/indicators, and sub-criteria of existing GBRTs and SATs were evaluated. Those tools that assess only the environmental sustainability of building projects were identified while those that attempt to encompass the triple dimension (economic, social, and environmental) were also identified and evaluated.

As the world's first GB assessment tool, the Building Research Establishment Environmental Assessment Method (BREEAM) is identified to be widely used. BREEAM considered ten categories under which the environmental performance of building projects is evaluated. The ten categories of BREEAM are Energy, Health and Well-being, Land Use and Ecology, Materials, Pollution, Transport, Management, Water, and Innovation (Reeder, 2010; Weerasinghe, 2012; CSI, 2013; BRE, 2018). The Leadership in Energy and Environmental Design (LEED) rating system is another widely used, and globally adapted tool aimed at reducing the negative environmental impacts of building projects (Uğur & Leblebici, 2018). LEED evaluates GB projects under nine categories of Indoor Environmental Quality, Energy and Atmosphere, Water Efficiency, Location and Transportation, Sustainable Sites, Integrative Process, Materials and Resources, Innovation, and Regional Priority (Wu et al., 2016; Ding et al., 2018). The Comprehensive Assessment System for Built Environment Efficiency (CASBEE) with international prominence and

DOI: 10.1201/9781003415961-14

regarded as the most studied rating tool after BREEAM and LEED (Li et al., 2017) evaluate the environmental performance of building projects based on six categories further classified as the environmental quality of building and environmental load reduction of the building. The six categories of CASBEE are Indoor Environment, Quality of Service, Outdoor Environment On-site, Energy, Resources and Materials, and Off-site Environment (Kawazu et al., 2005; JSBC, 2014).

There are other developed GBRTs or those adapted from BREEAM or LEED that do not thoroughly evaluate the sustainability performance of building projects. Based on this gap, GB councils, proponents, organisations, bodies, and agencies endeavoured to develop GBRTs or SATs that are well-adapted and suited for the country or region in which such a tool will be deployed. Examples of these tools include the Green Mark Scheme, the Green Rating for Integrated Habitat Assessment (GRIHA), and the Home Performance Index (HPI), amongst others. A perusal of these tools revealed that despite the attempt to cover the three dimensions of SD, they still do not comprehensively address SD issues.

However, there are records of other tools identified because of the quest to develop a region or country-specific SAT that holistically incorporates all dimensions of SD. Examples of such tools include the simplified method for evaluating building sustainability (SMEBS) in the early design phase for architects by Markelj et al. (2014), the healthcare building sustainability assessment method (HBSAtool-PT) for Portugal by Castro et al. (2017), the Building information modelling (BIM)-based Kazakhstan building sustainability assessment framework (KBSAF) by Akhanova et al. (2020), and the building sustainability assessment method (BSAM) for developing countries in sub-Saharan Africa by Olawumi et al. (2020). The SMEBS consists of ten evaluation categories, namely Pollution and Waste, Energy, Water Use, Materials, Sustainable Land Use, Wellbeing, Functionality, Technical Characteristics, Costs, and Property Value. The BSAM for developing countries in sub-Saharan Africa consists of eight categories, namely Sustainable Construction Practices, Site, and Ecology, Energy, Water, Material and Waste, Transportation, Indoor Environmental Quality, and Building Management. The BIM-based KBSAF consists of nine assessment categories, namely Construction Site Selection and Infrastructure, Building Architectural and Planning Solutions Quality, Indoor Environmental Quality and Comfort, Water Efficiency, Energy Efficiency, Green Building Materials, Waste, Economy, and Management. The HBSAtool-PT for Portugal consists of 22 assessment categories which are classified under sustainability areas of Environmental, Sociocultural and Functional, Economy, Technical, and Site.

Based on the analysis carried out by researchers globally on existing notable and widely used GBRTs and SATs, a similar trend of evaluation categories is identified that fall under the environmental aspect of SD (Bernardi et al., 2017; Yun et al., 2018; Díaz López et al., 2019). Some of the tools consist of the same evaluation categories while others either combine similar ones under a broader theme. A few of these mutual categories include Water, Energy, Indoor Environmental Quality, Materials and Resources, Land Use and Ecology/Site, and Management which all address environmental concerns. However, the recently developed GBRTs and

SATs by researchers are aimed at arriving at a tool that will equally address the environmental, social, and economic concerns of SD in a specific country or region. This is done by first identifying the gaps and addressing these concerns to arrive at a more integrated assessment tool.

Selection of Constructs for Biomimicry Sustainability Assessment Tool

To develop a more comprehensive assessment tool that addresses the SD demands in Africa and other developing countries coupled with the gaps identified in existing GBRTs and SATs, this book adopts the mutual categories/variables in line with reviewed literature and existing GBRTs and SATs. The determination of the constructs presented in this book is dependent on the mutual categories of these existing GBRTs and SATs, components of SD, and the sustainable development goals (SDGs) while drawing inspiration from nature to fill the gaps in the literature on assessment tools for building projects. Based on the precedents, this book considers ten broad categories to evaluate the sustainability performance of building projects which will also aid the achievement of SDGs and SD. The ten categories are classified into the three aspects of SD, namely Environmental aspect (Energy Efficiency, Water Efficiency, Materials Efficiency, Waste and Pollution, Site Ecology, and Resilience); Social Aspect (Collaboration, User Comfort, and Inclusivity and Equality); and Economic Aspect (Cost Management).

The conceptualised BioSAT is made up of the Energy Efficiency category which contains four criteria, Water Efficiency with three criteria, Materials Efficiency with four criteria, Waste and Pollution with four criteria, Site Ecology with three criteria, Resilience with five criteria, User Comfort with six criteria, Inclusivity and Equality with four criteria, Cost Management with nine criteria, and Collaboration with four criteria. The trio of resilience, inclusivity and equality, and collaboration are the identified gaps from the review of literature which were found to be peculiar to the developing nation's situation with Africa in focus (especially South Africa). Though these three identified variables have not been found in the existing GBRTs and SATs, they are deemed necessary to and influential in the sustainability performance of building projects and the SD of a nation. Table 10.1 presents a summary of the categories and criteria that have been conceptualised for the development of the BioSAT.

Latent Features of the Biomimicry Sustainability Assessment Tool

The section of this chapter presents a comprehensive breakdown of the ten evaluation categories that influence the sustainability performance of building projects. While three of the ten evaluation categories have been discussed extensively in the previous chapter, their inclusion in this section is to provide an outlook of their constituting criteria/indicators alongside others. This book, therefore, adopts the cogent categories peculiar to all notable GBRTs and SATS, namely, Energy Efficiency (EE) with four criteria, Water Efficiency (WE) with three criteria, Materials

Table 10.1 Conceptual framework latent features

Latent variable construct	Measurement variables
Energy Efficiency (EE)	EE1 – Efficient energy management
	EE2 – Renewable energy optimisation
	EE3 – Passive HVAC system
	EE4 – Use of energy-saving equipment
Water Efficiency (WE)	WE1 – Efficient water management
	WE2 – Use of water-efficient plumbing and sanitary fittings
	WE3 – Accessibility to safe and affordable water
Materials Efficiency (ME)	ME1 – Responsible materials processing
	ME2 – Use of locally sourced materials
	ME3 – Use of materials with verifiable eco-labels
	ME4 – Use of eco-friendly materials
Waste and Pollution (WP)	WP1 – Efficient waste management
	WP2 – Pollution management mechanisms
	WP3 – Reducing resource consumption
	WP4 – Design for disassembly and adaptability
Site Ecology (SE)	SE1 – Comprehensive site assessment
	SE2 – Responsible site spatial planning
	SE3 – Site ecological management
Resilience (RE)	RE1 – Effective disaster management plan
	RE2 – Design consideration of addressing natural disasters
	RE3 – Design consideration of addressing climate change
	RE4 – Use of resilient smart safety technologies
	RE5 – Optimising landscape elements for resilience
User Comfort (UC)	UC1 – Visual comfort of users
	UC2 – Effective indoor air quality management
	UC3 – Effective spatial acoustic performance
	UC4 – Proximity to basic amenities
	UC5 – Efficient user control and safety
	UC6 – Ease of building maintenance
Inclusivity and Equality (IE)	IE1 – Equitable user control
	IE2 – Spatial adaptability for user inclusiveness
	IE3 – Design consideration of cultural, ethnic, and religious differences of occupants
	IE4 – Consideration of community integration
Collaboration (CO)	CO1 – Effective multidisciplinary collaboration
	CO2 – Effective stakeholder conflict management
	CO3 – Effective collaboration with locals
	CO4 – Optimised corporate social responsibility
Cost Management (CM)	CM1 – Comprehensive project brief
	CM2 – Responsible construction practices
	CM3 – Engaging qualified and certified professionals
	CM4 – Consideration for building life cycle cost
	CM5 – Efficient technical management
	CM6 – Efficient adaptable space management
	CM7 – Use of cost-efficient materials and technologies
	CM8 – Consideration of commercial viability
	CM9 – Support for local contractors, suppliers, and businesses

Source: Author's compilation.

Efficiency (ME) with four criteria, Waste and Pollution (WP) with four criteria, Site Ecology (SE) with three criteria, User Comfort (UC) with six criteria, and Cost Management (CM) with nine criteria. However, the framework for this book includes Inclusivity and Equality (IE) with four criteria, Resilience (RE) with five criteria, and Collaboration with four criteria. The three added categories are the gaps in sustainability assessment frameworks found to be significant for the conceptualised BioSAT.

Energy Efficiency – EE

This category addresses energy efficiency in building projects. The step is aimed at ensuring control and reduction in demand and emissions to the barest minimum, and consumption of energy efficiently and responsibly. Globally, buildings and the built environment have been identified as energy-intensive owing to the numerous activities involved during the different lifecycles of a building (i.e., construction, operation, demolition). Therefore, to ensure a building project performs sustainably and optimally, energy efficiency is imperative, hence its inclusion in existing GBRTs and SATs. To this effect, certain criteria should be embraced and implemented in building projects to optimise their sustainability performance and eventual classification as a GB. A few of these strategies/mechanisms include the adoption of passive ventilation, cooling, lighting, and heating systems; the effective and optimised energy monitoring, metering, and performance; the utilisation of low carbon, non-emitting, and energy-saving fittings, appliances, and equipment; and the utilisation of clean and renewable energy.

Water Efficiency – WE

This category addresses the issue of water efficiency in building projects focusing on salient areas such as performance, use, cycle, and monitoring. As noted by the UN, developing countries are facing water challenges in terms of accessibility and availability of affordable, clean, and safe water sources. As a result, water efficiency is core in assessing the sustainability performance of building projects. This is also evident in the inclusion of this category in virtually all existing GBRTs and SATs in use globally. A few of the strategies/mechanisms adopted include effective rainwater, run-off water, wastewater, and greywater management; efficient water metering and monitoring; adopting of water leak detection and prevention systems; and the use of water efficient and quality plumbing and sanitary fittings, amongst others.

Materials Efficiency – ME

The materials efficiency category considers issues of efficiently reducing, using, reusing, and recycling resources and materials. This is aimed at mitigating the numerous environmental impacts of the utilisation of resources and materials during the building lifecycle. An examination of the various existing GBRTs and SATs revealed the predominance of this category, thereby justifying its inclusion in the conceptualised framework. This is because the materials and components of buildings are found to be significantly responsible for the unsustainable output of the built environment. A few of the strategies/mechanisms adopted under this category include responsible sourcing, harvesting, production, and use of materials/components; the use of reusable, recyclable, low-emitting, durable, locally sourced, and eco-labelled materials/components; and the use of materials/components free of toxins and red-list chemicals amongst others.

Waste and Pollution – WP

This category addresses the issue of waste and pollution throughout the lifecycle of building projects. This is aimed at combating the adverse environmental impacts of building projects because of waste and pollution generation. A few of the strategies/mechanisms adopted under this category include efficient and optimised management, disposal, collection, separation, storage, reduction, reuse, and recycling of waste; limiting and controlling of nuisance, pollution sources, and emission of harmful gases; reduced resource consumption; efficient flood and surface water management; and the safe disposal and handling of toxic waste amongst others.

Site Ecology – SE

This category addresses the impacts related to the design, planning, influence, and regeneration of site characteristics and biodiversity. A few of the strategies/mechanisms adopted under this category are long-term ecological maintenance and management; preservation of site ecological value and features; responsible site spatial planning, selection, assessment, and analysis; encouraging the use of brownfield sites; consideration for site security and environmental interactions; coordinated landscaping strategies and land conservation for human-scaled living; significant heat island reduction; and proper site layout to encourage eco-mobility and maximising land use for urban agriculture.

User Comfort – UC

This category is focused on the factors that ensure the well-being and comfort of occupants/users of the building projects with special attention to environmental ergonomics (i.e., acoustic comfort, visual comfort, and light quality). A few of the strategies/mechanisms adopted under this category include visual comfort; enhanced and effective indoor air quality assessment and management; thermal comfort; aesthetic consideration; quality of indoor and outdoor spaces; acoustic performance; safety and security of occupants/users; emission control; optimised natural lighting; comfortable artificial lighting; hygrothermal management; easy access to public transport and transportation routes; easy access to basic amenities; consideration for disabled occupants/users; and ease of use of low-emitting vehicles and other means of transportation.

Cost Management – CM

The cost management category entails the management and considerations for building lifecycle costs to ensure the sustainable performance of building projects. A few of the mechanisms adopted under this category include a comprehensive project brief and design; proper documentation for sustainable management; proper urban planning, design procedure, lifecycle cost, and service life planning; responsible construction practices and quality assurance of the project; design consideration for building certification; engagement of qualified and certified professionals; ensuring the shared cost of building; consideration of building location, buildability of design, quality building space, marketability, and commercial viability; effective cost monitoring mechanisms; consideration of professional fees; and reduced security and consumption cost.

Framework Specification and Rationalisation

Despite the popularity and proliferation of GBRTs and SATs, most of these tools still do not address the triple pillar of sustainability, a gap this book is intended to address by developing an SAT framework inspired by nature (biomimicry). The framework is based on existing and notable GBRTs, SATs, SD, and SDGs while drawing inspiration from the wholeness and attributes of natural systems known to be comprehensively sustainable. These factors, most of which have been subjected to continuous and thorough upgrades, consist of clear and comprehensive evaluation categories of sustainability assessment, rating, and certification framework.

Most of the notable existing GBRTs such as BREEAM, LEED, CASBEE, BEAM Plus, Green Star Australia, and Green Star SA, amongst numerous others, significantly consider the environmental aspect of SD. Prominent among the evaluation categories considered by these rating tools are energy, water, indoor environmental quality, management, materials and waste, pollution, land use and ecology, sustainable site, health and wellbeing, location and transportation, and innovation (which is aimed at recognising and rewarding added efforts to achieve sustainability in building projects). Other tools such as the Living Building Certification and Green Mark Scheme attempt to cover a bit more than the environmental aspect but are still unable to holistically address the tenets of SD. The Sustainable Building Assessment Tool developed by the Council for Scientific and Industrial Research (CSIR), the Simplified Method for Evaluating Building Sustainability (SMEBS) in the early design phase for architects, the Healthcare Building Sustainability Assessment Method (HBSAtool-PT) for Portugal, the Sustainability Assessment of High School Buildings (SAHSB) in Portugal, and the Building Sustainability Assessment Method (BSAM) for Developing Countries in sub-Saharan Africa are the few identified SATs that attempted to comprehensively evaluate buildings along the three dimensions of SD (Gibberd, 2003; Sebake, 2008; Markelj et al., 2014; Castro et al., 2017; Saraiva et al., 2019; Olawumi et al., 2020). However, the identified gaps of resilience, inclusivity and equality, and collaboration were not considered in any of the existing GBRTs, and SATs developed by GB bodies/organisations or postulated by research works of the authors earlier reviewed.

This book chapter seeks to project the analytical capability of these categories in the success of SATs to determine whether the realisation of SAT outcomes depends on the supposed attributes of the indicators, considering the necessities of the built environment and their sustainability performance as represented in other tools. The theorised framework conceptualises that SAT is founded on the correlation between the independent constructs which consist of the basic features through which the assessments are connected. The constructs established from the reviewed literature are acknowledged as core categories of an effective SAT. The established categories have been considered in the conceptualised BioSAT for the CI in developing countries (South Africa).

Structural Components of the Framework

The present conceptual framework hypothesis from which the sustainability assessment tool for the CI in South Africa (developing countries) comprises Energy Efficiency (EE), Water Efficiency (WE), Materials Efficiency (ME), Waste and Pollution (WP), Site Ecology (SE), Resilience (RE), User Comfort (UC), Inclusivity and Equality (IE), Cost Management (CM), and Collaboration (CO). The BioSAT to be examined in the postulated hypothesis is not dependent on previous research nor has it been tested but is a multidimensional structure consisting of EE, WE, ME, WP, SE, RE, UC, IE, CM, and CO. The conceptualised framework is presented schematically in Figure 10.1 (Framework 1.0).

As discussed above, the theoretical underpinning of this framework is derived from the analysis of existing GBRTs, SATs, SD tenets, SDGs, and biomimicry principles (drawing inspiration from nature on the wholeness and attributes of natural systems known to be comprehensively sustainable). The conceptualised framework is the knowledge that SAT for evaluating the sustainability performance of building projects is linked with the appraisal of the constructs highlighted earlier (i.e., EE, WE, ME, WP, SE, RE, UC, IE, CM, and CO). The comprehensiveness of the framework will be faulty and problematic without alluding to resilience, inclusivity and equality, and collaboration coupled with the addition of the extra independent constructs. In this book chapter, the appraisal of the SAT to determine the 'greenness' of building projects is assessed by measuring the actual condition of the GB industry which is an exogenous variable in the framework.

Measurement Components of the Framework

The measurement components of the hypothesised framework consist of the ensuing BioSAT categories/factors, namely: EE (4 measurement variables/criteria); WE (3 measurement variables/criteria); ME (4 measurement variables/criteria); and WP (4 measurement variables/criteria). Others include SE (3 measurement variables/criteria); RE (5 measurement variables/criteria); UC (6 measurement variables/criteria); IE (4 measurement variables/criteria); CM (9 measurement variables/criteria);

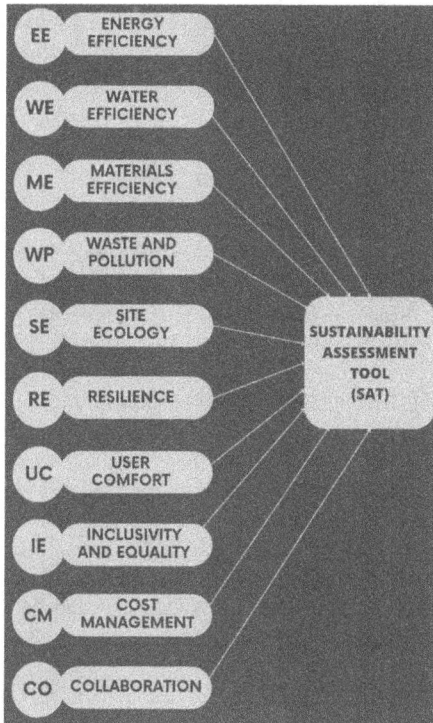

Figure 10.1 Conceptualised framework of the biomimicry sustainability assessment tool

and CO (4 measurement variables/criteria). The factors for the sustainable performance of building projects in developing countries (South Africa) using the BioSAT are postulated in the current framework.

Summary

This chapter reviewed and discussed the conceptual framework upon which the study is underpinned. The conceptualised BioSAT is presented in this chapter as inferred from the rigorous and in-depth review of existing GBRTs, and SATs as presented in the early part of this book. Seven categories namely, Energy Efficiency (EE), Water Efficiency (WE), Materials Efficiency (ME), Waste and Pollution (WP), Site Ecology (SE), User Comfort (UC), and Cost Management (CM) were adopted from the reviewed GBRTs and SATs as the main constructs. The conceptualised BioSAT postulated a multidimensional structure composed of ten latent variables, including Energy Efficiency (EE), Water Efficiency (WE), Materials Efficiency (ME), Waste and Pollution (WP), Site Ecology (SE), User Comfort (UC), Cost Management (CM), Collaboration (CO), Inclusivity and Equality (IE), and Resilience (RE) all rooted and in tandem with SD tenets, the SDGs, and biomimicry principles. These ten latent variables are what constitute the critical success factors for improving the sustainability performance of buildings. The next chapter focuses on presenting a perspective on the future of BioSAT vis a vis emerging technologies and construction innovations.

References

Akhanova, G., Nadeem, A., Kim, J. R., & Azhar, S. (2020). A multi-criteria decision-making framework for building sustainability assessment in Kazakhstan. *Sustainable Cities and Society, 52*, 1–11.

Bernardi, E., Carlucci, S., Cornaro, C., & Bohne, R. (2017). An analysis of the most adopted rating systems for assessing the environmental impact of buildings. *Sustainability, 9*(1226), 1–27.

Building Research Establishment. (2018). *BREEAM.* Retrieved April 03, 2018, from https://www.bre.co.uk/page.jsp?id=829

Castro, M. F., Mateus, R., & Bragança, L. (2017). Development of a healthcare building sustainability assessment method – Proposed structure and system of weights for the Portuguese context. *Journal of Cleaner Production, 148*, 555–570.

Construction Specifications Institute (2013). *The CSI sustainable design and construction practice guide.* Somerset: John Wiley & Sons.

Díaz López, C., Carpio, M., Martín-Morales, M., & Zamorano, M. (2019). A comparative analysis of sustainable building assessment methods. *Sustainable Cities and Society, 49*, 1–22.

Ding, Z., Fan, Z., Tam, V. W. Y., Bian, Y., Li, S., Illankoon, I. M., Chethana S, & Moon, S. (2018). Green building evaluation system implementation. *Building and Environment, 133*, 32–40.

Gibberd, J. (2003). *Building systems to support sustainable development in developing countries.* Pretoria, South Africa: CSIR Building and Construction Technology.

Japan Sustainable Building Consortium (2014). *CASBEE for building (new construction): Technical manual.* Japan: Institute for Building Environment and Energy Conservation.

Kawazu, Y., Shimada, N., Yokoo, N., & Oka, T. (2005). *Comparison of the assessment results of BREEAM, LEED, GBTool and CASBEE.* World Sustainable Building Conference (SB05Tokyo), 1700–1705.

Li, Y., Chen, X., Wang, X., Xu, Y., & Chen, P. (2017). A review of studies on green building assessment methods by comparative analysis. *Energy and Buildings, 146*, 152–159.

Markelj, J., Kitek Kuzman, M., Grošelj, P., & Zbašnik-Senegačnik, M. (2014). A simplified method for evaluating building sustainability in the early design phase for architects. *Sustainability, 6*(12), 8775–8795.

Olawumi, T. O., Chan, D. W. M., Chan, A. P. C., & Wong, J. K. W. (2020). Development of a building sustainability assessment method (BSAM) for developing countries in sub-Saharan Africa. *Journal of Cleaner Production, 263*, 1–37.

Reeder, L. (2010). *Guide to green building rating systems: Understanding LEED, green globes, Energy Star, the national green building standard, and more.* Hoboken, New Jersey: John Wiley & Sons.

Saraiva, T. S., Almeida, M., & Bragança, L. (2019). Adaptation of the SBTool for sustainability assessment of high school buildings in Portugal – SAHSBPT. *Applied Sciences, 9*(13), 1–15.

Sebake, T. N. (2008). Review of appropriateness of international environmental assessment tools for a developing country. *World Sustainable Building Conference,* 1–7.

Uğur, L. O., & Leblebici, N. (2018). An examination of the LEED green building certification system in terms of construction costs. *Renewable and Sustainable Energy Reviews, 81*, 1476–1483.

Weerasinghe, U. G. D. (2012). *Development of a framework to assess sustainability of building projects*. Canada: University of Calgary.

Wu, P., Mao, C., Wang, J., Wang, X., & Song, Y. (2016). A decade review of the credits obtained by LEED v2.2 certified green building projects. *Building and Environment, 102*, 167–178.

Yun, Y., Cho, D., & Chae, C. (2018). Analysis of green building certification system for developing g-SEED. *Future Cities and Environment, 4*(1), 1–9.

11 Validation of the Biomimicry Sustainability Assessment Tool

Overview of the Analytical Hierarchy Process

In the quest to identify According to Saaty (1990), the analytical hierarchy process (AHP) is mathematically involved 'with scaling problem and what sort of numbers to use, and how to correctly combine the priorities resulting from them'. Developed in 1980 by Thomas L. Saaty, the AHP is a mathematical tool/model based on a hierarchical structure for managing quantitative (QUAN) and qualitative (QUAL) multi-criteria elements involved in decision-making (DM) behaviour (Taherdoost, 2017). The development of AHP emanated from the experience gained by Saaty while leading research within the United States Arms Control and Disarmament Agency (Bhushan & Rai, 2004). It is a technique for solving complicated DM problems with the use of mathematics and psychology (Singh, 2016). AHP equips the DM process with the requisite framework to solve complex problems while enabling the decision maker to make effective decisions by first breaking down the complex problems into elementary variables, sequentially organising these variables, and lastly, ascribing a numerical value to evaluate the importance of each variable (Khaireddin, 2016). The higher the level of AHP, the broader the decision, with the topmost elements being decomposed into categories, criteria, and attributes (Sarkis et al., 2009). In this technique, a multi-level hierarchical structure of objectives (categories/indicators), criteria, sub-criteria, and alternatives are utilised (Triantaphyllon & Mann, 1995) while proffering an efficient way of addressing multidimensional DM (Castro et al., 2017).

Two options exist with the use of AHP for measurement, namely relative and absolute. Relative measurements are employed to compare criteria in all problems (pairwise comparison) using a fundamental scale of absolute numbers, while absolute measurement is used to rank alternatives in terms of the criteria and sum up to obtain an overall rating on a ratio scale (Saaty, 1986; Saaty, 1989). As indicated by Goepel (2018), AHP is based on pairwise comparison inputs, and the weights are computed by locating the assertive (dominant) eigenvector (EV) of a positive reciprocal decision matrix. According to Hamouda et al. (2012), the AHP process allows for using aggregation of sub-categories/criteria of indicators/categories influencing the decision as it prevents the inclusion of indicators/categories adjudged as irrelevant by different decision makers. Described by Kotykhov (2005) as a

DOI: 10.1201/9781003415961-15

research methodology, AHP can be employed to structure, measure and synthesise elements/factors (variables) that affect DM. By constructing a pairwise comparison matrix, AHP is used to determine the consistency of weightings for criteria while providing an appropriate way of managing imprecision criteria and integrating QUAL analysis with QUAN factors (Abdulhasan et al., 2019).

According to Saaty (1989), AHP as a descriptive theory 'helps to explore the complexity of the decision honestly and without oversimplification to express judgements according to preference intensity, and finally to derive a solution that clearly and rigorously captures these intensities according to magnitude'. Based on the study of Lee (1998), the advantages of AHP are summarised under the following categories:

- Unity: AHP provides a single, easily understood, flexible model for a wide range of unstructured problems.
- Process Repetition: AHP enables DMs to refine their definition of a problem and to improve their judgement and understanding through repetition.
- Complexity: AHP integrates systematic approaches to solving complex problems.
- Judgement and Consensus: AHP does not insist on consensus but synthesises a representative outcome from diverse judgements.
- Decomposition: AHP can decompose large problems into small ones which can easily be understood.
- Trade-offs: AHP takes into consideration the relative priorities of factors in a system and enables DMs to select the best alternative based on their goals.
- Hierarchic Structuring: AHP reflects the natural tendency of the mind to sort elements of a system into different levels and to group like elements in each level.
- Synthesis: AHP leads to an overall estimate of the desirability of each alternative.
- Measurement: AHP provides a scale for measuring intangibles and a method for establishing priorities.
- Consistency: AHP tracks the logical consistency of judgements used in determining priorities.

Despite the strengths of the AHP technique, a few weaknesses/disadvantages are identified with its use, a situation that is also peculiar to every method/technique of research. However, the strengths/advantages have seen its application in numerous DM situations such as Choice (selection of an alternative from a set of alternatives); Prioritisation/Evaluation (determination of the relative advantage of a set of alternatives); Resource allocation (finding the appropriate combination of alternatives subject to a variety of constraints); Benchmarking (of systems/processes with other known systems/processes); and Quality management (Bhushan & Rai, 2004). Hence, AHP is applied in this research study to determine and allocate the weight of the importance of the parameters that constitute the conceptualised BioSAT for the construction industry (CI) in developing countries using South Africa as a case study. Table 11.1 presents a

Table 11.1 AHP application in previous studies on sustainability assessment

Author(s) and year published	Topic of study	AHP applications
Alawneh et al. (2019)	A Novel Framework for Integrating United Nations Sustainable Development Goals into Sustainable Non-residential Building Assessment and Management in Jordan	AHP and Delphi technique
Markelj et al. (2014)	A Simplified Method for Evaluating Building Sustainability in the Early Design Phase for Architects	AHP technique
Castro et al. (2017)	Development of a Healthcare Building Sustainability Assessment Method – Proposed Structure and System of Weights for the Portuguese Context	AHP technique
Huang and Hsu (2011)	Framework development for state-level appraisal indicators of sustainable construction	Max-Min Fuzzy Delphi method and Fuzzy AHP
AbdelAzim et al. (2017)	Development of an energy efficiency rating system for existing buildings using Analytic Hierarchy Process – The case of Egypt	AHP technique
Cappai et al. (2018)	The Integration of Socio-Economic Indicators in the CASBEE-UD Evaluation System: A Case Study	AHP technique and GIS
Sarkis et al. (2009)	A Sustainability Decision Model for the Built Environment	AHP and ANP technique
Koo (2007)	Development of sustainability assessment model for underground infrastructure	AHP and WSM technique
Jayawickrama (2014)	Conceptual Framework for Environmental Rating Systems for Infrastructure Projects in Sri Lanka: Application to Small Hydropower Projects	AHP technique and Sample Group (interview)
Banani et al. (2016)	The Development of Building Assessment Criteria Framework for Sustainable Non-residential Buildings in Saudi Arabia	AHP technique

Source: Author's compilation

few of the identified studies that utilised AHP as a DM technique or tool in the built environment with a focus on sustainability assessment. These provided credence to the choice and imperativeness of AHP for developing the conceptualised tool.

To employ AHP, Saaty (1990) and Al Barqouni (2015) listed six steps to follow, namely define the problem, develop the AHP hierarchy, perform the pairwise comparison, estimate the relative weights, check the consistency, and obtain the overall rating. The studies of Mu and Pereyra-Rojas (2018) comprehensively highlighted the following steps to be followed in using AHP:

- Develop a model for the decision (breakdown the decision into a hierarchy of goals, criteria, and alternatives);
- Derive priorities (weights) for the criteria which are compared pairwise concerning the desired goal to derive their weights, and check the consistency of judgements (judgement review to ensure a reasonable level of consistency in terms of transitivity and proportionality);
- Derive local priorities (preferences) for the alternatives concerning each criterion separately by following a similar process in the previous step, while checking and adjusting the consistency as required;
- Derive overall priorities (model synthesis) by combining all obtained alternative priorities as a weighted sum to consider each criterion's weight and to establish the overall priorities of the alternatives, while the alternative with the highest overall priority becomes the best choice;
- Perform sensitivity analysis (which studies how changes in the weights of the criteria could affect the result to understand the rationale behind the results obtained); and
- Make a final decision based on the synthesised results and sensitivity analysis.

Summarising the steps involved in the AHP technique, Tzeng and Huang (2011) highlighted four steps as follows: Problem decomposition into a hierarchy of interrelated elements and setting up the hierarchy system; comparison of the comparative weight between the attributes of the decision elements to establish the reciprocal matrix; synthetisation of the individual subjective judgement estimation of the relative weight; and aggregation of the relative weights of the elements to determine the best strategies or alternatives.

Analysis and Synthesis Using Analytic Hierarchy Process

For the assessment of the conceptualised biomimicry sustainability assessment tool (BioSAT) for the CI in developing countries (South Africa), a total of 46 criteria were identified. These criteria were subsets of the ten categories (energy efficiency, water efficiency, materials efficiency, waste and pollution, site ecology, resilience, user comfort, inclusivity and equality, cost management, and collaboration) which are grouped into three dimensions of environmental, economic, and social. The questionnaire comprised pairwise comparisons of the individual parameters (dimensions, categories, and criteria) on the same hierarchy level. In total, the questionnaire contained 107 pairwise comparisons that were determined by the experts (decision-makers). The experts used the provided Saaty number scale of 1 to 9 to allocate judgements in pairwise comparisons. Invited experts have the task of assigning values while performing the pairwise comparison of the elements. The adapted fundamental scale table also described as the ratio scale in the AHP (Tzeng & Huang, 2011) or pairwise comparison scale (Mu & Pereyra-Rojas, 2018) or scale

Table 11.2 AHP scale of relative importance

Intensity of importance	Verbal judgement/Definition	Explanation
1	Equally Important	Two activities contribute equally to the objective
3	Moderately Important	Experience and judgement slightly favour one activity over another
5	Strongly Important	Experience and judgement strongly favour one activity over another
7	Very Strongly Important	An activity is favoured very strongly over another; its dominance demonstrated in practice
9	Extremely Important	The evidence favouring one activity over another is of the highest possible order of affirmation
2, 4, 6, 8	Intermediate Values	When compromise is needed
Reciprocals of above	If activity *i* has one of the above non-zero numbers assigned to it when compared with activity *j*, then *j* has the reciprocal value when compared with *i*	A reasonable assumption

Source: Saaty and Vargas (2012).

of relative importance (Triantaphyllou & Mann, 1995) is presented in Table 11.2. Super Decisions v3 software was used for the analysis, synthesis, and justification of inputs made by the experts. The AHP method in this present study was executed based on the summarised steps in the following order: build a hierarchy, make comparisons, calculate weights, check consistency, and produce results. Figure 11.1 shows the hierarchy built for the conceptualised BioSAT framework and subsequently utilised for the AHP analysis.

Local and Global Weightings of Dimensions, Categories, and Criteria of the Framework

Making decisions as it is in the case of sustainability assessment tools (SATs) requires the allocation of weights and values to established parameters. In green or sustainable buildings' evaluation, the term weighting of criteria is used to describe the relative importance of such a parameter within its group (Bhatt & Macwan, 2012). According to Aragon (2017), science-based, objective measurements and methods should be employed in the development of weightings and yardsticks for parameters, categories, and criteria. Owing to its intuitiveness and mathematical rigour, AHP has become one of the most widely utilised multi-criteria decision-making (MCDM) methods which allows the inclusion of intangibles (intuition, subjective preferences, and experiences) in a structured and logical way (Mu & Pereyra-Rojas, 2018). In making a final decision, which is the final stage of AHP

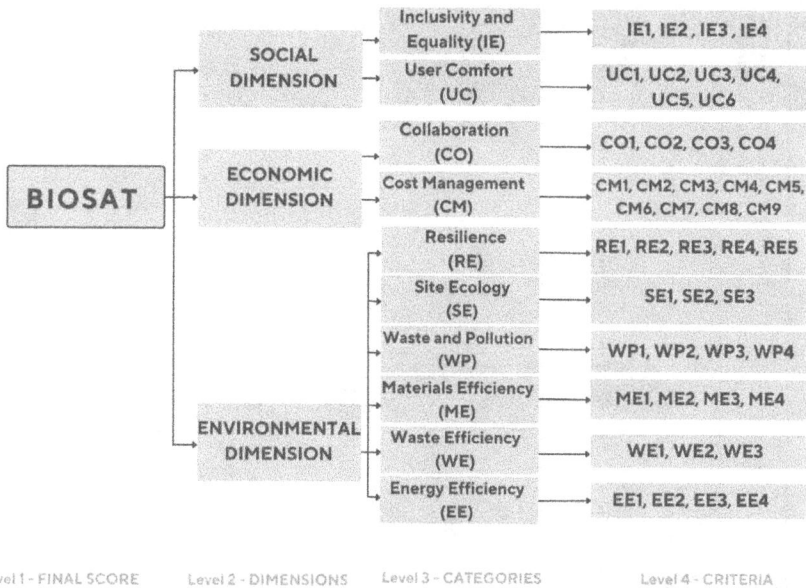

Figure 11.1 Hierarchy of the conceptualised biomimicry sustainability assessment tool

analysis, priorities, or weightings (local and global) are generated, thereby making the technique one of the best in determining the relative importance of elements within a hierarchy. By generating the overall priorities, the model synthesis of the problem decomposed to form the AHP hierarchy is achieved (Mu & Pereyra-Rojas, 2018).

Weightings are used to determine the priority or significance of each aspect, category, criterion, and sub-criterion within an SAT (Banani et al., 2016). Within a group of parameters, the sum of LW which is also the priority weighting is one. However, the global weight (GW) for a parameter within a group is derived by multiplying LW by hierarchically higher parameters. According to Markelj et al. (2014), GW is derived mathematically using the equation below:

$$GW_{P,i} = LW_{A,j} \times LW_{CA,k} \times LW_{CR,i} \tag{11.1}$$

where GWP = global weight of parameter; A = aspect; CA = category; CR = criteria that represent the level on the hierarchical structure; LW = local weight; i, j, k indexes = 1, …, n representing the individual criteria, category, or aspect at each hierarchical level. The total of the LW within an individual group of the category parameters will be 1 while the GW in each hierarchy will also be 1 when summed up (Bhatt & Macwan, 2012). Super Decisions v3 software was used to compute and generate the global weightings. Table 11.3 presents the results of weights allocation to the dimensions, categories, and criteria of the BioSAT hierarchy.

Table 11.3 Results of weights allocation to dimensions, categories, and criteria of BioSAT

Dimensions/Local weight	Local weight	Categories	Local weight	Criteria	Global weight	Global weight	Global weight	Global rank
ENVIRONMENTAL	0.1726	Energy Efficiency (EE)	0.2979	Efficient energy management (EE1)	0.0192	0.0646	0.3739	21
			0.2435	Renewable energy optimisation (EE2)	0.0157			33
			0.1945	Passive HVAC system (EE3)	0.0126			39
			0.2642	Use of energy-saving equipment (EE4)	0.0171			28
	0.1194	Water Efficiency (WE)	0.3337	Efficient water management (WE1)	0.0149	0.0446		34
			0.2492	Use of water-efficient plumbing and sanitary fittings (WE2)	0.0111			45
			0.4171	Accessibility to safe and affordable water (WE3)	0.0186			22
	0.2569	Materials Efficiency (ME)	0.2259	Responsible materials processing (ME1)	0.0217	0.0961		17
			0.2506	Use of locally sourced materials (ME2)	0.0241			13
			0.2284	Use of materials with verifiable eco-labels (ME3)	0.0219			16
	0.1756	Waste and Pollution (WP)	0.2951	Use of eco-friendly materials (ME4)	0.0284	0.0657		11
			0.2632	Efficient waste management (WP1)	0.0173			26
			0.2067	Pollution management mechanisms (WP2)	0.0136			37
			0.2682	Reducing resource consumption (WP3)	0.0176			25
			0.2619	Design for disassembly and adaptability (WP4)	0.0172			27
	0.1072	Site Ecology (SE)	0.3296	Comprehensive site assessment (SE1)	0.0132	0.0401		38
			0.3614	Responsible site spatial planning (SE2)	0.0145			35
			0.3090	Site ecological management (SE3)	0.0124			41
	0.1683	Resilience (RE)	0.1778	Effective disaster management plan (RE1)	0.0112	0.0629		44
			0.1799	Design consideration for addressing natural disasters (RE2)	0.0113			43
			0.1831	Design consideration for addressing climate change (RE3)	0.0115			42
			0.2601	Use of resilient smart safety technologies (RE4)	0.0164			30
			0.1991	Optimising landscape elements for resilience (RE5)	0.0125			40

(Continued)

Table 11.3 (Continued)

Dimensions/Local weight	Local weight	Categories	Local weight	Criteria	Global weight	Global weight	Global weight	Global rank
SOCIAL	0.5228	User Comfort (UC)	0.1418	Visual comfort of users (UC1)	0.0234	0.1650	0.3157	14
			0.2339	Effective indoor air quality management (UC2)	0.0386			4
			0.1387	Effective spatial acoustic performance (UC3)	0.0229			15
			0.1502	Proximity to basic amenities (UC4)	0.0248			12
			0.2079	Efficient user control and safety (UC5)	0.0343			10
			0.1276	Ease of building maintenance (UC6)	0.0211			18
	0.4772	Inclusivity and Equality (IE)	0.2379	Equitable user control (IE1)	0.0358	0.1507		9
			0.2465	Spatial adaptability for user inclusiveness (IE2)	0.0371			8
			0.2688	Design consideration for cultural, ethnic, and religious differences of occupants (IE3)	0.0405			3
			0.2468	Consideration for community integration (IE4)	0.0372			7

(Continued)

Table 11.3 (Continued)

Dimensions/Local weight	Local weight	Categories	Local weight	Criteria	Global weight	Global weight	Global weight	Global rank
ECONOMIC	0.4807	Cost Management (CM)	0.0696	Comprehensive project brief (CM1)	0.0104	0.1492	0.3104	46
			0.1096	Responsible construction practices (CM2)	0.0164			31
			0.1136	Engaging qualified and certified professionals (CM3)	0.0170			29
			0.1306	Consideration for building life cycle cost (CM4)	0.0195			20
			0.1338	Efficient technical management (CM5)	0.0200			19
			0.0947	Efficient adaptable space management (CM6)	0.0141			36
			0.1200	Use of cost-efficient materials and technologies (CM7)	0.0179			24
			0.1211	Consideration for commercial viability (CM8)	0.0181			23
			0.1072	Support for local contractors, suppliers, and businesses (CM9)	0.0160			32
	0.5194	Collaboration (CO)	0.2315	Effective multidisciplinary collaboration (CO1)	0.0373	0.1612		6
			0.2790	Effective stakeholders conflict management (CO2)	0.0450			1
			0.2535	Effective collaboration with locals (CO3)	0.0409			2
			0.2360	Optimised corporate social responsibility (CO4)	0.0381			5

Table 11.4 Consistency ratio of the dimensions, categories, and criteria of BioSAT

Item	Elements	Consistency ratio (CR)
Dimension	Environmental, Social, and Economic	**0.01**
Category	**Environmental**	**0.05**
Criteria	Energy Efficiency	0.03
	Water Efficiency	0.00
	Materials Efficiency	0.01
	Waste and Pollution	0.02
	Site Ecology	0.00
	Resilience	0.04
Category	**Social**	**0.00**
Criteria	User Comfort	0.04
	Inclusivity and Equality	0.03
Category	**Economic**	**0.00**
Criteria	Cost Management	0.03
	Collaboration	0.01

Consistency of Judgements

The value of consistency ratio (CR) for all the elements in the hierarchy is presented in Table 11.4. From the table, it is evident that the CR values for all the judgements of the experts (decision makers) are below the 0.1 or 10% acceptable standard. Hence, the judgements of the experts (decision makers) are deemed consistent, acceptable, and valid for the present research study (Vargas, 2010; Tzeng & Huang, 2011; Banani et al., 2016; Taherdoost, 2017; Mu & Pereyra-Rojas, 2018).

Discussion of the Pairwise Comparison Questionnaire Results

This section focuses on discussing descriptive statistics, and the AHP analysis. The descriptive statistics were based on the demographic/background information of the respondents (experts/decision makers) which constitute the first section of the pairwise comparison questionnaire survey.

Discussion of the Expert's Background Information

This section discusses the background of the respondents (experts/decision makers) based on their educational qualification, profession, professional affiliation, operational base, employer, years of experience in the CI, number of green building projects executed, and number of green-rated/certified projects executed. Table 11.5 presents the experts' demographic characteristics. From Table 11.5, the first observation made showed that most of the respondents are master's degree holders (65.8%), followed by those with a doctoral degree (28.9%) while only 5.3% hold a bachelor/honour's degree. This result is an indication that the respondents possess the requisite qualifications. Also, 26.3% of the respondents are architects, 18.4% are quantity surveyors, 13.2% are construction managers and construction project managers respectively, and 5.3% are civil engineers,

Table 11.5 Experts' demographic characteristics

Characteristics	Description	Frequency (n)	Percentage (%)
Educational Qualification	Bachelor/Honour's Degree	2	5.3
	Master's Degree	25	65.8
	PhD/Doctorate Degree	11	28.9
	Total	**38**	**100**
Profession	Architect	10	26.3
	Quantity Surveyor	7	18.4
	Civil Engineer	2	5.3
	Industrial Engineer	1	2.6
	Electrical Engineer	2	5.3
	Mechanical Engineer	2	5.3
	Construction Manager	5	13.2
	Construction Project Manager	5	13.2
	Project Manager	2	5.3
	Town/Urban Planner	2	5.3
	Total	**38**	**100**
Professional Affiliation	Biomimicry South Africa	2	5.3
	Green Building Council South Africa (GBCSA)	5	13.2
	South African Council for the Architectural Profession (SACAP)	5	13.2
	South African Council for the Quantity Surveying Profession (SACQSP)	7	18.4
	Engineering Council of South Africa (ECSA)	6	15.8
	South African Council for Planners (SACPLAN)	2	5.3
	Chartered Institute of Building (CIOB)	5	13.2
	South African Council for the Project and Construction Management Professions (SACPCMP)	6	15.8
	Total	**38**	**100**
Office Location/ Operational Base	Gauteng	22	57.9
	Free State	2	5.3
	Western Cape	14	36.8
	Total	**38**	**100**
Employer	Public sector	11	28.9
	Private sector	24	63.2
	Public and Private sector	3	7.9
	Total	**38**	**100**
Years of Experience	6–10 years	8	21.1
	11–15 years	22	57.9
	16–20 years	8	21.1
	Total	**38**	**100**
Number of Green Building projects executed	3–4 projects	5	13.2
	5–6 projects	17	44.7
	7–8 projects	12	31.6
	9–10 projects	3	7.9
	More than 10 projects	1	2.6
	Total	**38**	**100**
Number of Green rated/certified projects executed	1–2 projects	12	31.6
	3–4 projects	25	65.8
	5–6 projects	1	2.6
	Total	**38**	**100**

electrical engineers, mechanical engineers, project managers and town/urban planners respectively while 2.6% represents the only industrial engineer respondent. This presents a fair distribution of the professionals in the CI with an indication of their level of involvement in project execution. From Table 11.5, all the respondents are active members of their respective professional bodies duly certified and recognised by the South African government. A total of 57.9% of the respondents (expert decision makers) have their main office or operational base located in Gauteng Province, 36.8% are from the Western Cape Province and 5.3% from the Free State Province. This result reflects the fact that most infrastructural and green-certified projects are in the Gauteng and Western Cape provinces with a scattered concentration of these projects in other provinces of South Africa (Simpeh & Smallwood, 2015; Windapo & Goulding, 2015; GBCSA, 2018; Masia et al., 2020).

Also from Table 11.5, 63.2% of the respondents work or are employed by private entities, 28.9% by public entities, and 7.9% work for public and private entities. A total of 57.9% of the respondents have 11–15 years of working experience in the CI, while 21.1% have 6–10 years and 16–20 years respectively. This result is an indication that the experts/decision-makers possess ample years of experience required for participating in the research study. Furthermore, 44.7% of the respondents have been involved in 5–6 green building projects, 31.6% in 7–8 green building projects, 13.2% in 3–4 green building projects, 7.9% in 9–10 green building projects and 2.6% in more than 10 green building projects. Also, 65.8% of the respondents have been involved in 3–4 green-rated/certified building projects, 31.6% in 1–2 green-rated/certified building projects and 2.6% in 5–6 green-rated/certified building projects. The study by Windapo (2014) informed of the paucity of green building projects in South Africa compared to the major markets of the United States, Europe, Asia, UAE, and Australia. Considering the involvement of the respondents in a reasonable number of green and certified building projects, it can be inferred that they are well-qualified and suitable to participate in the research study.

Discussion of the Model Results Using AHP

Super Decisions v3 software was used for AHP to test and evaluate the importance of the ten-factor constructs of energy efficiency (EE), water efficiency (WE), materials efficiency (ME), waste and pollution (WP), site ecology (SE), resilience (RE), user comfort (UC), inclusivity and equality (IE), cost management (CM), and collaboration (CO). The test was premised on the conjecture that an effective sustainability assessment tool (SAT) is significantly influenced by EE, WE, ME, WP, SE, RE, UC, IE, CM, and CO factors. Since the experts (respondents) who participated (defining weights using AHP) in the research study all work in South Africa, their judgements can be considered a realistic reflection of sustainability priorities and demands of the regional context.

Energy Efficiency and the Conceptualised Biomimicry Sustainability
Assessment Tool

The energy efficiency (EE) category contributes 6% of the total GW, ranking as the 3rd most important category within the environmental dimensions. For the variables (EE1, EE2, EE3, and EE4) of the category, they respectively contribute 2%, 2%, 1%, and 2% of the total GW of the goal (BioSAT) while respectively ranking as the 21st, 33rd, 39th, and 28th most important criteria that influence the sustainability performance of BioSAT. On the other hand, the sensitivity analysis showed the relative importance of the weightings with the adjustment in the experts' judgement suggesting the significance of the variables. For the sensitivity analysis, EE category still contributes 6% while the variables/criteria (EE1, EE2, EE3, and EE4) respectively contribute 2%, 1%, 1%, and 2% of the total GW of the goal (BioSAT) while respectively ranking as the 26th, 34th, 39th, and 31st most important criteria that influence the sustainability performance of BioSAT.

Building environmental assessment and labelling tools are identified as one of the effective and potent ways of improving building sustainability performance (Cole & Jose Valdebenito, 2013). According to Lee and Burnett (2008), LEED and BREEAM are the two most representative building environmental assessment systems globally. Having the issue of energy as a standalone area of measured performance in both tools and numerous others (Adegbile, 2013; Olawumi et al., 2020) indicates the significance and influence in achieving sustainable building (SB) projects. The result is in tandem with the studies of Roderick et al. (2009), Markelj et al. (2014), Malek and Grierson (2016), and Oguntona et al. (2019) which confirms the significance of energy efficiency in an assessment tool for SBs. Addressing the energy aspect of building projects will also help in getting rid of the negative image of the CI as an energy-intensive sector. Therefore, the energy efficiency category is highly influential in an effective BioSAT for building projects. It can therefore be argued that energy efficiency has a significant relationship with the sustainability performance of building projects.

Water Efficiency and the Conceptualised Biomimicry Sustainability
Assessment Tool

The water efficiency (WE) category contributes 4% of the total GW, ranking as the 5th most important category within the environmental dimensions. For the variables (WE1, WE2, and WE3) of the category, they respectively contribute 1%, 1%, and 2% of the total GW of the goal (BioSAT) while respectively ranking as the 34th, 45th, and 22nd most important criteria that influence the sustainability performance of BioSAT. On the other hand, the sensitivity analysis showed the relative importance of the weightings with the adjustment in the experts' judgement suggesting the significance of the variables. For the sensitivity analysis and similar to the AHP result, the WE category still contributes 4% while the variables/criteria (WE1, WE2, and WE3) respectively contribute 1%, 1%, and 2% of the total GW of the goal (BioSAT) while respectively ranking as the 35th, 46th, and 27th most important criteria that influence the sustainability performance of BioSAT.

According to Bhatt and Macwan (2012), one of the nine major goals of SB is water efficiency achieved by reduced water use and wastewater generation. This is evident in most building assessment tools where environmental categories such as water efficiency are in common agreement amongst the tools (Banani et al., 2016). The result is in tandem with the studies of Drejeris and Kavolynas (2014), and Díaz López et al. (2019) indicating the significance and influence of water efficiency in SATs for achieving SB projects. By addressing the water aspect of building projects, the negative environmental footprint of the CI as a resource-intensive sector will be mitigated. Therefore, the water efficiency category is highly influential in an effective BioSAT for building projects. It can therefore be argued that water efficiency has a significant relationship with the sustainability performance of building projects.

Materials Efficiency and the Conceptualised Biomimicry Sustainability Assessment Tool

The materials efficiency (ME) category contributes 10% of the total GW, ranking as the most important category within the environmental dimensions. For the variables (ME1, ME2, ME3 and ME4) of the category, they respectively contribute 2%, 2%, 2% and 3% of the total GW of the goal (BioSAT) while respectively ranking as the 17th, 13th, 16th and 11th most important criteria that influence the sustainability performance of BioSAT. On the other hand, the sensitivity analysis showed the relative importance of the weightings with the adjustment in the experts' judgement suggesting the significance of the variables. For the sensitivity analysis, ME category contributes 9% while the variables/criteria (ME1, ME2, ME3 and ME4) respectively contribute 2%, 2%, 2%, and 3% of the total GW of the goal (BioSAT) while respectively ranking as the 21st, 16th, 19th, and 12th most important criteria that influence the sustainability performance of BioSAT.

In the quest for solutions that reduce environmental degradation, materials efficiency (low-energy consuming materials, reusable waste materials, materials requiring less financial investment) has been identified as a cogent strategy (Bawankule & Rajurkar, 2017). Attesting to its importance in achieving SBs, significant credit weighting is placed on the standalone materials or resources category in the Green Star SA rating system, as it is almost impossible to address any building without considering its resource or material aspects (van Reenen, 2014). The presence of materials or resource categories in almost all of the GBRTs and SATs is another pointer to the imperativeness of the aspect in achieving SB. The result is in tandem with the studies of Thiel et al. (2013), Ferwati et al. (2019) and Okokpujie et al. (2020) which indicate the importance of materials efficiency in enhancing the environmental performance of buildings. By being mindful of the material use or resource use in building projects, the negative environmental footprint of the CI as a resource-intensive and high waste-generating sector will be addressed. Therefore, the materials efficiency category is highly influential in an effective BioSAT for building projects. It can therefore be argued that materials efficiency has a significant relationship with the sustainability performance of building projects.

Waste and Pollution and the Conceptualised Biomimicry Sustainability Assessment Tool

The waste and pollution (WP) category contributes 7% of the total GW, ranking as the 2nd most important category within the environmental dimensions. For the variables (WP1, WP2, WP3, and WP4) of the category, they respectively contribute 2%, 1%, 2%, and 2% of the total GW of the goal (BioSAT) while respectively ranking as the 26th, 37th, 25th, and 27th most important criteria that influence the sustainability performance of BioSAT. On the other hand, the sensitivity analysis showed the relative importance of the weightings with the adjustment in the experts' judgement suggesting the significance of the variables. For the sensitivity analysis, the WP category contributes 9% while the variables/criteria (WP1, WP2, WP3, and WP4) respectively contribute 2%, 1%, 2%, and 2% of the total GW of the goal (BioSAT) while respectively ranking as the 29th, 37th, 28th, and 30th most important criteria that influence the sustainability performance of BioSAT.

As defined by Li (2013), construction waste is a mix of surplus materials emanating from road, demolition, renovation, refurbishment, construction, excavation, and site clearance works. Coupled with pollution, the duo of waste and pollution are identified through literature as one of the major negative impacts of the CI on the environment. Hence, the presence in almost all GBRTs and SATs is an evaluation category for enhancing the performance of SBs. The result is in tandem with the studies by Awadh (2017) and Varma and Palaniappan (2019) which indicate the influence of waste and pollution in SATs for enhancing the environmental performance of buildings. By considering the waste and pollution aspect of building projects, the negative environmental footprint of the CI will be significantly reduced. Therefore, the waste and pollution category is highly influential in an effective BioSAT for building projects. It can therefore be argued that waste and pollution have a significant relationship with the sustainability performance of building projects.

Site Ecology and the Conceptualised Biomimicry Sustainability Assessment Tool

The site ecology (SE) category contributes 4% of the total GW, ranking as the 6th (last) most important category within the environmental dimensions. For the variables (SE1, SE2, and SE3) of the category, they respectively contribute 1%, 1%, and 1% of the total GW of the goal (BioSAT) while respectively ranking as the 38th, 35th, and 41st most important criteria that influence the sustainability performance of BioSAT. On the other hand, the sensitivity analysis showed the relative importance of the weightings with the adjustment in the expert's judgement suggesting the significance of the variables. For the sensitivity analysis and just like the AHP result, the SE category still contributes 4% while the variables/criteria (SE1, SE2, and SE3) respectively still contribute 1%, 1%, and 1% of the total GW of the goal

(BioSAT) while respectively ranking as the 38th, 36th, and 42nd most important criteria that influence the sustainability performance of BioSAT.

While some tools such as BREEAM and Green Star Australia address the category as 'land use and ecology', LEED as 'sustainable sites' while SBTool refers to the category as 'site selection, project planning and development' (Sev, 2011; Doan et al., 2017). It is therefore important to note that different building assessment schemes/tools use the same terminology for different entities or different terminologies to describe the same entity (Lee, 2013). Despite the different wording regarding the category by researchers and existing tools, the target and intent of its evaluation are still the same as with 'site ecology' as used in this research study. The study agrees with most of the existing GBRTs and SATs and studies indicating the influence of site ecology on SATs for enhancing the environmental performance of buildings (Jawali & Fernández-Solís, 2008; Ameen et al., 2015). By considering the site ecology aspect of building projects, the negative environmental footprint of the CI will be significantly reduced while preserving biodiversity and the natural ecosystem. Therefore, the site ecology category is highly influential in an effective BioSAT for building projects. It can therefore be argued that site ecology has a significant relationship with the sustainability performance of building projects.

Resilience and the Conceptualised Biomimicry Sustainability Assessment Tool

The resilience (RE) category contributes 6% of the total GW, ranking as the 4th most important category within the environmental dimensions. For the variables (RE1, RE2, RE3, RE4, and RE5) of the category, they respectively contribute 1%, 1%, 1%, 2%, and 1% of the total GW of the goal (BioSAT) while respectively ranking as the 44th, 43rd, 42nd, 30th, and 40th most important criteria that influence the sustainability performance of BioSAT. On the other hand, the sensitivity analysis showed the relative importance of the weightings with the adjustment in the experts' judgement suggesting the significance of the variables. For the sensitivity analysis, the RE category contributes 6% while the variables/criteria (RE1, RE2, RE3, RE4, and RE5) respectively contribute 1%, 1%, 1%, 1%, and 1% of the total GW of the goal (BioSAT) while respectively ranking as the 45th, 44th, 43rd, 33rd and 40th most important criteria that influence the sustainability performance of BioSAT.

According to Markelj et al. (2014), the development, adaptation and use of SATs must consider climatic, cultural, geographical, political, and economic situations in the country or region. With the process of globalisation, climate change and natural disasters posing grave threats to the sustainability of the built environment (Scott, 2017), incorporating elements of resilience into SATs will ensure SBs are achieved. The novel integration of the resilience category in this research study aligns with the global clamour for a resilient built environment and SD as evidenced in SDGs 9 (build *resilient* infrastructure, promote inclusive and sustainable industrialisation and foster innovation) and 11 (make cities and human settlements inclusive, safe, *resilient* and sustainable) of the UN. By considering the resilience aspect in building projects, the constituting elements and systems of buildings will be able to withstand events while optimising the sustainability performance of such elements and systems before, during, and after the event. Therefore, the resilience category is highly influential in an effective BioSAT for building projects. It can therefore be argued that resilience has a significant relationship with the sustainability performance of building projects.

User Comfort and the Conceptualised Biomimicry Sustainability Assessment Tool

The user comfort (UC) category contributes 17% of the total GW, ranking as the most important category within the social dimensions. For the variables (UC1, UC2, UC3, UC4, UC5, and UC6) of the category, they respectively contribute 2%, 4%, 2%, 2%, 3%, and 2% of the total GW of the goal (BioSAT) while respectively ranking as the 14th, 4th, 15th, 12th, 10th, and 18th most important criteria that influence the sustainability performance of BioSAT. On the other hand, the sensitivity analysis showed the relative importance of the weightings with the adjustment in the experts' judgement suggesting the significance of the variables. For the sensitivity analysis, the UC category still contributes 17% while the variables/criteria (UC1, UC2, UC3, UC4, UC5, and UC6) respectively contribute 2%, 4%, 2%, 3%, 4%, and 2% of the total GW of the goal (BioSAT) while respectively ranking as the 13th, 5th, 14th, 11th, 10th, and 15th most important criteria that influence the sustainability performance of BioSAT.

The study agrees with most of the existing GBRTs, SATs and studies indicating the significance of user comfort (also referred to as indoor environment in some tools) in SATs for enhancing the sustainability performance of building projects (Xing et al., 2009; Markelj et al., 2013; Li et al., 2016; Oh et al., 2017). Also, the study of Kim et al. (2013) emphasises that SBs could benefit from a people-centred approach by focusing on user needs, health and satisfaction. By considering the user comfort aspect in building projects, the health and comfort of users of SBs will be able to withstand events while optimising the sustainability performance of such elements and systems before, during, and after the event. Therefore, the user comfort category is highly influential in an effective BioSAT for building projects. It can therefore be argued that user comfort has a significant relationship with the sustainability performance of building projects.

Inclusivity and Equality and the Conceptualised Biomimicry Sustainability Assessment Tool

The inclusivity and equality (IE) category contributes 15% of the total GW, ranking as the 2nd most important category within the social dimensions. For the variables (IE1, IE2, IE3, and IE4) of the category, they respectively contribute 4%, 4%, 4%, and 4% of the total GW of the goal (BioSAT) while respectively ranking as the 9th, 8th, 3rd, and 7th most important criteria that influence the sustainability performance of BioSAT. On the other hand, the sensitivity analysis showed the relative importance of the weightings with the adjustment in the experts' judgement suggesting the significance of the variables. For the sensitivity analysis, the IE category contributes 16% while the variables/criteria (IE1, IE2, IE3, and IE4) respectively still contribute 4%, 4%, 4%, and 4% of the total GW of the goal (BioSAT) while respectively ranking as the 9th, 8th, 3rd, and 7th most important criteria that influence the sustainability performance of BioSAT.

According to Segerstedt and Abrahamsson (2019), social sustainability involves desirable, democratic, connected, diverse, and equitable qualities that enhance and result in a good quality of life. However, the social aspect of SD has received low attention when compared to the environment in the development of assessment tools for buildings. SATs that claim to integrate the three dimensions of SD are also found not to consider the influence of inclusivity and equality which is a cogent factor of SD as postulated by the UN. As a novel part of this research study, the inclusion of inclusivity and equality aligns and conforms with the SDGs 4, 5, 8, 9, 10, 11, and 16 which reflect the importance of inclusivity and equality in realising the goal of SD (UN, 2015; Villeneuve et al., 2017). By considering the inclusivity and equality aspect in building projects, the goal of achieving an all-encompassing SD agenda will be ably realised while optimising the potential and opportunities availed through an effective SAT. Therefore, the inclusivity and equality category is highly influential in an effective BioSAT for building projects. It can therefore be argued that inclusivity and equality have a significant relationship with the sustainability performance of building projects.

Cost Management and the Conceptualised Biomimicry Sustainability Assessment Tool

The cost management (CM) category contributes 15% of the total GW, ranking as the 2nd most important category within the economic dimensions. For the variables (CM1, CM2, CM3, CM4, CM5, CM6, CM7, CM8, and CM9) of the category, they respectively contribute 1%, 2%, 2%, 2%, 2%, 1%, 2%, 2%, and 2% of the total GW of the goal (BioSAT) while respectively ranking as the 46th, 31st, 29th, 20th, 19th, 36th, 24th, 23rd, and 32nd most important criteria that influence the sustainability performance of BioSAT. On the other hand, the sensitivity analysis showed the relative importance of the weightings with the adjustment in the experts' judgement suggesting the significance of the variables. For the sensitivity analysis, the CM category still contributes 15% while the variables/criteria (CM1, CM2, CM3, CM4, CM5, CM6, CM7, CM8 and CM9) respectively contribute 1%, 2%, 2%, 2%, 2%, 2%, 2%, 2%, and 2% of the total GW of the goal (BioSAT) while respectively ranking as the 41st, 24th, 23rd, 18th, 17th, 32nd, 22nd, 20th, and 25th most important criteria that influence the sustainability performance of BioSAT.

The CM category entails the management of and considerations for building life-cycle costs to ensure the sustainable performance of building projects. While tools such as BREEAM and Green Star Australia use the term 'management' and DGNB the term 'economic quality and quality of planning', SBtool uses the term 'cost and economic' to address the economic aspect of SD in SB evaluation (Chehrzad et al., 2016). The result is in tandem with the studies of Sarkis et al. (2009), Bryce et al.

(2017), Markelj et al. (2014) and renowned tools such as BREEAM, DGNB, and Green Star Australia which iterate the significance of cost management category in an assessment tool for SBs. Therefore, the cost management category is highly influential in an effective BioSAT for building projects. It can therefore be argued that cost management has a significant relationship with the sustainability performance of building projects.

Collaboration and the Conceptualised Biomimicry Sustainability Assessment Tool

The collaboration (CO) category contributes 16% of the total GW, ranking as the 3rd most important category within the economic dimensions. For the variables (CO1, CO2, CO3, and CO4) of the category, they respectively contribute 4%, 5%, 4%, and 4% of the total GW of the goal (BioSAT) while respectively ranking as the 6th, 1st, 2nd, and 5th most important criteria that influence the sustainability performance of BioSAT. On the other hand, the sensitivity analysis showed the relative importance of the weightings with the adjustment in the experts' judgement suggesting the significance of the variables. For the sensitivity analysis, the CO category contributes 17% while the variables/criteria (CO1, CO2, CO3, and CO4) respectively contribute 4%, 5%, 4%, and 4% of the total GW of the goal (BioSAT) while ranking as the 6th, 1st, 2nd, and 4th most important criteria that influence the sustainability performance of BioSAT.

According to Böggemann and Both (2014), collaboration is the only means of creating novel approaches to optimising growth and efficiency in ways that address the global challenges facing our already resource-constrained world. Hence, the identification of collaborative partnerships is a core determinant in the implementation of the UN 2030 SDGs. Missing within existing GBRTs and SATs, the novel integration of the collaboration category as a standalone in this research study aligns with SDG 17 which is aimed at strengthening the means of implementation and revitalisation of the global partnership of SD. By embracing and emulating the outstanding collaborative processes in the natural ecosystem which has made them exhibit sustainable outputs, achieving a true SB will be a reality in the CI. Therefore, the collaboration category is highly influential in an effective BioSAT for building projects. It can therefore be argued that collaboration has a significant relationship with the sustainability performance of building projects.

Extent the Conceptualised Tool Fits the Identified Constructs

The findings from AHP and the sensitivity analysis suggest that all the ten identified variables had a direct positive influence in determining the effectiveness and sustainability performance of SAT for building projects. The assumption that an effective BioSAT is a product of the direct influence of the exogenous variables in predicting the sustainability performance of building projects in the construction industry of developing countries using South Africa as a case study is sustained. The findings support previous research studies, existing GBRTs and SATs, and UN

SDGs which informed that an effective building assessment tool is a product of balanced integration of the three dimensions of SD (Markelj et al., 2013; Markelj et al., 2014; UN, 2015; Olawumi & Chan, 2018; Olawumi et al., 2020).

The postulated relationships between the dimensions, categories and criteria were found to be statistically significant and reliably valid. The result suggested that all the factors adequately measure the overall goal of the hierarchy which is BioSAT. The relationship between the sustainability performance of BioSAT and User Comfort was found to be the most significant, closely followed by the Collaboration, Inclusivity and Equality categories. For the latter two categories to be identified as vital ahead of the common factors in existing and renowned GBRTs and SATs, it is evident that there is a consensus that the novelty in this current research study which integrates the Collaboration, and Inclusivity and Equality categories will result in more SBs. Similarly, the Resilience category of BioSAT according to the findings of the study will lead to more and truly SBs in South Africa. Therefore, the sustainable performance of building projects can be achieved in developing countries (South Africa) through the developed BioSAT which considers Energy Efficiency, Water Efficiency, Materials Efficiency, Waste and Pollution, Site Ecology, Resilience, User Comfort, Inclusivity and Equality, Cost Management, and Collaboration evaluation categories.

Summary of Analytical Hierarchical Process for Biomimicry Sustainability Assessment Tool

The conceptualised biomimicry sustainability assessment tool (BioSAT) framework is based on building sustainability assessment and the acquired relative importance weights for parameters from AHP analysis and synthesis. For all the judgement matrices involved, the CR was well below 0.1 or 10% making the results consistent, reliable, and valid. In the AHP hierarchy, the criteria on level 4, categories on level 3, and dimensions on level 2 all have a significant influence on the realisation of the goal (BioSAT) which is the final output on level 1. It is therefore concluded that 3 dimensions, 10 categories and 46 criteria as schematically represented in Figure 11.2 adequately represent the Biomimicry Sustainability Assessment Tool (BioSAT) for developing countries (South Africa). The figure also shows the final weight that each dimension, category and criterion contributes to influencing the overall sustainability performance of the framework. The sum of the weight of each element in a level must be 100% with each quota represented by the weighting.

Summary

This chapter focuses on the validation of the constructs of the proposed sustainability assessment tool by adopting the analytical hierarchical process (AHP) using South Africa as a case study. The chapter comprehensively described an overview of the AHP in sustainability assessment tool research. The chapter also presented the computed local and global weightings of the dimensions, categories, and criteria

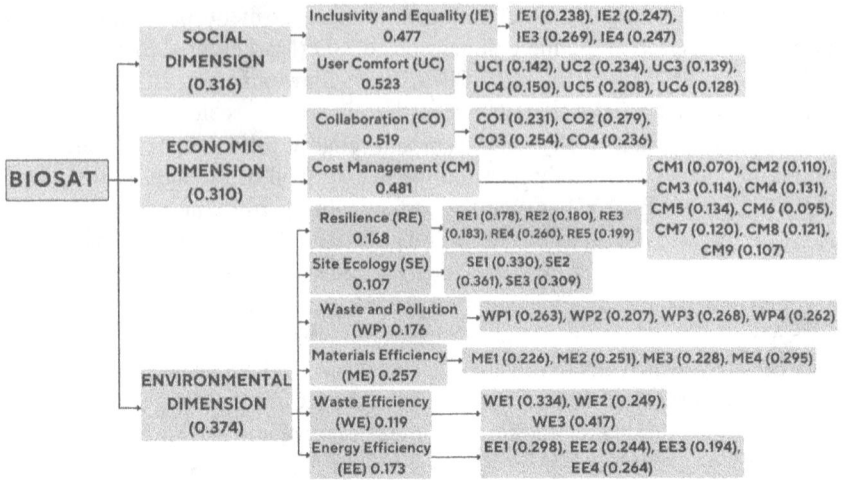

Figure 11.2 AHP framework for the biomimicry sustainability assessment tool

of the conceptualised tool from the AHP analysis. Also, the chapter discussed the composition of the expert (decision-makers) panel, the demographics of the experts, and the results of the model using AHP. Based on the pairwise comparison questionnaire results, this chapter discussed the findings and supported them from the literature. The judgements of the experts (decision makers) are deemed consistent, acceptable, and valid for all the ten constructs thereby improving the reliability of this study. The AHP results played a significant role in determining the various dimensions, categories, and criteria that contribute to the sustainable performance of sustainability assessment tools in South Africa. These constructs led to the development of the biomimicry sustainability assessment tool for the construction industry in South Africa.

References

AbdelAzim, A. I., Ibrahim, A. M., & Aboul-Zahab, E. M. (2017). Development of an energy efficiency rating system for existing buildings using analytic hierarchy process – The case of Egypt. *Renewable and Sustainable Energy Reviews, 71*, 414–425.

Abdulhasan, M. J., Hanafiah, M. M., Satchet, M. S., Abdulaali, H. S., Toriman, M. E., & Al-Raad, A. A. (2019). Combining GIS, fuzzy logic and AHP models for solid waste disposal site selection in Nasiriyah, Iraq. *Applied Ecology and Environmental Research, 17*(3), 6701–6722.

Adegbile, M. B. (2013). Assessment and adaptation of an appropriate green building rating system for Nigeria. *Journal of Environment and Earth Science, 3*(1), 1–10.

Al Barqouni, H. M. (2015). *Application of analytic hierarchy process (AHP) in risk assessment for construction building projects*. Gaza: The Islamic University.

Alawneh, R., Ghazali, F., Ali, H., & Sadullah, A. F. (2019). A novel framework for integrating United Nations Sustainable Development Goals into sustainable non-residential building assessment and management in Jordan. *Sustainable Cities and Society, 49*, 1–20.

Ameen, R. F. M., Mourshed, M., & Li, H. (2015). A critical review of environmental assessment tools for sustainable urban design. *Environmental Impact Assessment Review, 55*, 110–125.

Aragon, T. J. (2017). *Deriving criteria weights for health decision making: A brief tutorial.* San Francisco, USA: San Francisco Department of Public Health.

Awadh, O. (2017). Sustainability and green building rating systems: LEED, BREEAM, GSAS and Estidama critical analysis. *Journal of Building Engineering, 11*, 25–29.

Banani, R., Vahdati, M. M., Shahrestani, M., & Clements-Croome, D. (2016). The development of building assessment criteria framework for sustainable non-residential buildings in Saudi Arabia. *Sustainable Cities and Society, 26*, 289–305.

Bawankule, M. V., & Rajurkar, V. J. (2017). Suitability assessment of fly-ash GGBS-based rammed earth material – A sustainable construction building material – A review. *Journal of Geotechnical Studies, 2*(2), 1–9.

Bhatt, R., & Macwan, J. E. M. (2012). Global weights of parameters for sustainable buildings from consultants' perspectives in Indian context. *Journal of Architectural Engineering, 18*(3), 233–241.

Bhushan, N., & Rai, K. (2004). *Strategic decision making: Applying the analytic hierarchy process.* United States of America: Springer-Verlag London.

Böggemann, M., & Both, N. (2014). The resilience action initiative: An introduction. In R. Kupers (Ed.), *Turbulence* (pp. 23–34). Netherlands: Amsterdam University Press.

Bryce, J., Brodie, S., Parry, T., & Lo Presti, D. (2017). A systematic assessment of road pavement sustainability through a review of rating tools. *Resources, Conservation & Recycling, 120*, 108–118.

Cappai, F., Forgues, D., & Glaus, M. (2018). The integration of socio-economic indicators in the CASBEE-UD evaluation system: A case study. *Urban Science, 2*(1), 28.

Castro, M. F., Mateus, R., & Bragança, L. (2017). Development of a healthcare building sustainability assessment method – Proposed structure and system of weights for the Portuguese context. *Journal of Cleaner Production, 148*, 555–570.

Chehrzad, M., Pooshideh, S. M., Hosseini, A., & Sardroud, J. M. (2016). A review on green building assessment tools: Rating, calculation and decision-making. *WIT Transactions on Ecology and the Environment, 204*, 397–404.

Cole, R. J., & Jose Valdebenito, M. (2013). The importation of building environmental certification systems: International usages of BREEAM and LEED. *Building Research & Information, 41*(6), 662–676.

Díaz López, C., Carpio, M., Martín-Morales, M., & Zamorano, M. (2019). A comparative analysis of sustainable building assessment methods. *Sustainable Cities and Society, 49*, 1–22.

Doan, D. T., Ghaffarianhoseini, A., Naismith, N., Zhang, T., Ghaffarianhoseini, A., & Tookey, J. (2017). A critical comparison of green building rating systems. *Building and Environment, 123*, 243–260.

Drejeris, R., & Kavolynas, A. (2014). Multi-criteria evaluation of building sustainability behavior. *Procedia - Social and Behavioral Sciences, 110*, 502–511.

Ferwati, M. S., AlSuwaidi, M., Shafaghat, A., & Keyvanfar, A. (2019). Employing biomimicry in urban metamorphosis seeking for sustainability: Case studies. *Architecture, City and Environment (ACE), 14*(40), 133–162.

Goepel, K. D. (2018). Implementation of an online software tool for the analytic hierarchy process (AHP-OS). *International Journal of the Analytic Hierarchy Process, 10*(3), 469–487.

Green Building Council South Africa (2018). *Integrated annual report 2018.* South Africa: GBCSA.

Hamouda, M. A., Anderson, W. B., & Huck, P. M. (2012). Employing multi-criteria decision analysis to select sustainable point-of-use and point-of-entry water treatment systems. *Water Science and Technology: Water Supply, 12*(5), 637–647.

Huang, R., & Hsu, W. (2011). Framework development for state-level appraisal indicators of sustainable construction. *Civil Engineering and Environmental Systems, 28*(2), 143–164.

Jawali, R., & Fernández-Solís, J. L. (2008). A building sustainability rating index (BSRI) for building construction. *Proceedings of the 8th International Post Graduate Research Conference,* 1–16.

Jayawickrama, T. S. (2014). *Conceptual framework for environmental rating systems for infrastructure projects in Sri Lanka: Application to small hydropower projects.* National University of Singapore.

Khaireddin, M. A. (2016). Building and applying a proposed model for suppliers' selection using multi-criteria approach: The analytic hierarchy process (AHP). *International Journal of Statistics and Systems, 11*(1), 47–65.

Kim, M. J., Oh, M. W., & Kim, J. T. (2013). A method for evaluating the performance of green buildings with a focus on user experience. *Energy and Buildings, 66,* 203–210.

Koo, D. H. (2007). *Development of sustainability assessment model for underground infrastructure.* Arizona State University.

Kotykhov, M. (2005). *Determinant attributes of customer choice of banks, supplying mortgage products.* Auckland University of Technology.

Lee, S. (1998). *Analytic hierarchy approach for transport project appraisal: An application to Korea.* University of Leeds.

Lee, W. L. (2013). A comprehensive review of metrics of building environmental assessment schemes. *Energy and Buildings, 62,* 403–413.

Lee, W. L., & Burnett, J. (2008). Benchmarking energy use assessment of HK-BEAM, BREEAM and LEED. *Building and Environment, 43*(11), 1882–1891.

Li, Y. (2013). *Developing a sustainable construction waste estimation and management system.* Hong Kong University of Science and Technology.

Li, B., Li, Y., Yu, W., & Yao, R. (2016). A multidimensional model for green building assessment: A case study of a highest-rated project in Chongqing. *Energy & Buildings, 125,* 231–243.

Malek, S., & Grierson, D. (2016). A contextual framework for the development of a building sustainability assessment method for Iran. *Open House International, 41*(2), 64–75.

Markelj, J., Kitek Kuzman, M., Grošelj, P., & Zbašnik-Senegačnik, M. (2014). A simplified method for evaluating building sustainability in the early design phase for architects. *Sustainability, 6*(12), 8775–8795.

Markelj, J., Kitek Kuzman, M., & Zbašnik-Senegačnik, M. (2013). A review of building sustainability assessment methods. *Archit. Res, 1,* 22–31.

Masia, T., Kajimo-Shakantu, K., & Opawole, A. (2020). A case study on the implementation of green building construction in Gauteng province, South Africa. *Management of Environmental Quality: An International Journal, 31*(3), 602–623.

Mu, E., & Pereyra-Rojas, M. (2018). *Practical decision making using super decisions v3: An introduction to the analytic hierarchy process.* Switzerland: Springer International Publishing AG.

Oguntona, O. A., Maseko, B. M., Aigbavboa, C. O., & Thwala, W. D. (2019). Barriers to retrofitting buildings for energy efficiency in South Africa. *IOP Conference Series. Materials Science and Engineering, 640,* 1–7.

Oh, O., Lim, J., Lim, C., & Kim, S. (2017). A health performance and cost optimization model for sustainable healthy buildings. *Journal of Asian Architecture and Building Engineering, 16*(2), 303–309.

Okokpujie, I. P., Okonkwo, U. C., Bolu, C. A., Ohunakin, O. S., Agboola, M. G., & Atayero, A. A. (2020). Implementation of multi-criteria decision method for selection of suitable material for development of horizontal wind turbine blade for sustainable energy generation. *Heliyon, 6*(1), 1–10.

Olawumi, T. O., & Chan, D. W. (2018). A scientometric review of global research on sustainability and sustainable development. *Journal of Cleaner Production, 183*, 231–250.

Olawumi, T. O., Chan, D. W. M., Chan, A. P. C., & Wong, J. K. W. (2020). Development of a building sustainability assessment method (BSAM) for developing countries in sub-Saharan Africa. *Journal of Cleaner Production, 263*, 1–37.

Roderick, Y., McEwan, D., Wheatley, C., & Alonso, C. (2009). Comparison of energy performance assessment between LEED, BREEAM and green star. *Eleventh International IBPSA Conference*, 1167–1176.

Saaty, T. L. (1986). Absolute and relative measurement with The AHP. The most livable cities in The United States. *Socio-Economic Planning Sciences, 20*(6), 327–331.

Saaty, T. L. (1989). Decision-making, scaling, and number crunching. *Decision Sciences, 20*(2), 404–409.

Saaty, T. L. (1990). How to make a decision: The analytic hierarchy process. *European Journal of Operational Research, 48*(1), 9–26.

Saaty, T. L., & Vargas, L. G. (2012). Models, methods, concepts & applications of the analytic hierarchy process. In F. S. Hillier (Ed.), *International series in operations research & management science* (2nd ed.). New York, USA: Springer Science & Business Media.

Sarkis, J., Meade, L., & Presley, A. (2009). *A sustainability decision model for the built environment*. Worcester, MA, USA: George Perkins Marsh Institute, Clark University.

Scott, L. (2017). *Lessons in resilience: From biological systems to human food systems*. US: Oberlin College and Conservatory.

Segerstedt, E., & Abrahamsson, L. (2019). Diversity of livelihoods and social sustainability in established mining communities. *The Extractive Industries and Society, 6*(2), 610–619.

Sev, A. (2011). A comparative analysis of building environmental assessment tools and suggestions for regional adaptations. *Civil Engineering and Environmental Systems, 28*(3), 231–245.

Simpeh, E. K., & Smallwood, J. J. (2015). Factors influencing the growth of green building in the South African construction industry. *Smart and Sustainable Built Environment (SASBE) Conference 2015*, 311–320.

Singh, B. (2016). Analytical hierarchical process (AHP) and fuzzy AHP applications – A review paper. *International Journal of Pharmacy and Technology, 8*(4), 4925–4946.

Taherdoost, H. (2017). Decision making using the analytic hierarchy process (AHP): A step-by-step approach. *International Journal of Economics and Management Systems, 2*, 244–246.

Thiel, C., Campion, N., Landis, A., Jones, A., Schaefer, L., & Bilec, M. (2013). A materials life cycle assessment of a net-zero energy building. *Energies, 6*(2), 1125–1141.

Triantaphyllou, E., & Mann, S. H. (1995). Using the analytic hierarchy process for decision making in engineering applications: Some challenges. *International Journal of Industrial Engineering: Applications and Practice, 2*(1), 35–44.

Tzeng, G., & Huang, J. (2011). *Multiple attribute decision making: Methods and applications*. Florida, USA: CRC Press.

United Nations. (2015). *Transforming our world: The 2030 Agenda for Sustainable Development*. Retrieved 24 September 2019 from https://sustainabledevelopment.un.org/post2015/transformingourworld

van Reenen, C. A. (2014). *Principles of material choice with reference to the Green Star SA rating system*. SA: Alive2Green.

Vargas, R. V. (2010). Using the analytic hierarchy process (AHP) to select and prioritize projects in a portfolio. *PMI Global Congress 2010 – North America*, 1–22.

Varma, C. R. S., & Palaniappan, S. (2019). Comparison of green building rating schemes used in North America, Europe and Asia. *Habitat International, 89*, 1–13.

Villeneuve, C., Tremblay, D., Riffon, O., Lanmafankpotin, G., & Bouchard, S. (2017). A systemic tool and process for sustainability assessment. *Sustainability, 9*(10), 1–29.

Windapo, A. O. (2014). Examination of green building drivers in the South African construction industry: Economics versus ecology. *Sustainability, 6*(9), 6088–6106.

Windapo, A. O., & Goulding, J. S. (2015). Understanding the gap between green building practice and legislation requirements in South Africa. *Smart and Sustainable Built Environment, 4*(1), 67–96.

Xing, Y., Horner, R. M. W., El-Haram, M. A., & Bebbington, J. (2009). A framework model for assessing sustainability impacts of urban development. *International Conference on Whole Life Urban Sustainability and Its Assessment, 33*(3), 1–23.

Index

For Product Safety Concerns and Information please contact our EU
representative GPSR@taylorandfrancis.com
Taylor & Francis Verlag GmbH, Kaufingerstraße 24, 80331 München, Germany